Scanner Modification Handbook

Volume 1

By Bill Cheek

CRB Research Books, Inc.

P. O. Box 56
COMMACK, NEW YORK 11725

Dedicated to

My loving wife and children, "Sunbunny, Beach Bunny and Li'l Grizz," who, with clenched jaws, furrowed brows, and silent tongues, tolerated the many weeks of my single-minded dedication.

"Doctor Rigormortis"

Third Printing

ISBN 0-939780-11-9

Realistic is a Registered Trademark of the Radio Shack Division of Tandy Corporation. Bearcat and Regency are Registered Trademarks of the Uniden Corporation of America.

Cover and interior design and layout by Graphics Department of CRB Research Books, Inc. Special thanks to Robin Smith.

Copyright (C) 1990 by CRB Research Books, Inc. All rights reserved. Reproduction of the content or any portion thereof in any manner whatsoever without the express written permission of copyright owner is prohibited. Any violations will be vigorously prosecuted to the fullest extent of the federal copyright laws. Neither the publisher nor the author assume any liability with respect to the use of the information contained herein. Any modifications made to equipment are the sole decision and complete responsibility of the owner of the equipment. Use of all equipment must be in complete accordance with any and all federal, state, and local laws, regulations, and ordinances in effect.

Contents

Forward by Tom Kneitel	7
1. Scanning & The Law	**10**
Communications Act of 1934	10
Electronics Communications Privacy Act of 1986	12
Modifications to Your Scanner	13
2. Scanner Hobby Tips, Hints & Kinks	**17**
Antennas	17
Height	18
Gain & Bandwidth	18
Choosing a Scanner Antenna	19
TV Antennas for Scanners	22
Installing Antennas	23
Constructing & Modifying Antennas	23
Radio Shack Discone Low Band Improvement	23
Replacement Antennas for Handheld Scanners	25
An All-Occasion Portable Scanner Antenna	25
An Emergency or Indoor Antenna (2 Methods)	27
Coaxial Cable (Feedline) & Connectors	29
Comparison of Coaxial Cables	29
db Loss to % Loss Conversion	31
Coaxial Cable Recommendations for Scanner Use	32
Connectors & Splices	32
Preamplifiers & Converters	33
Tips & Hints on Scanner Receivers	35
Evaluating & Choosing a Scanner	35
Scanner Features & Specifications	36
Emergency Power for Your Scanner	37
Analysis of PRO-2004 Power Requirements	37
Make a DC Power Cable for the PRO-2004/2005	39
Emergency Power for Other Scanners	40
Battery Eliminators for Handheld Scanners	41
Emergency 12-14 VDC Power for Handheld Scanners	42
Tips on NiCd Batteries	43
Tips on Alkaline Batteries	44
An Alkaline Battery Power Source for Handhelds	44
Operating Hints & Techniques	44
Increasing Your Knowledge	44
Frequency Directories	45
Hobby Periodical Literature	45
Radio Clubs	46
Building Your Scanner Library	46
Organizing Your Monitoring Station	47
Scanner Frequency & Channel Management	48
Practical Program for 400 Channels	49
Master Programming Plan for 400-Channel Scanners	50
40-Channel Bank Memory Detail	53
6,400 Channel Memory Overview	54
Scanner Operations	55
Instant Weather Channel Access	55

    Receiving Out-of-Band Frequencies   55
    Monitoring Quiet & Seldom Used Frequencies   55
        Tape Recording With a VOX
        Tape Recording With a Tape Recorder Switch
    Grounding for Safety & Performance   57

3. Cellular Mobile Telephone: Explained   61
    CMT's & The ECPA   61
    The Cellular Concept   62
    Sequence of Events in a Cellular Telephone Call   62
    CMT Frequency Allocations   63
    CMT Computer Operations   63
    Typical Cellular System   64
    How Authorized Personnel Might Monitor a CMT System   67
    Wireline Company Cell Transmit Frequencies   68
    Non-Wireline Company Cell Transmit Frequencies   69

4. Performance Improvement Modifications for Scanners   73
    Introduction, History & Background   73
    Why The PRO-2004 & PRO-2005 Are so Prominent   74
    General Approach to Modifications   75
        Adding New Frequency Bands   76
        Scan & Search Speed Increases   78
        Adding S-Meters   78
        Adding Tape Recorder Functions   79
        Automatic Tape Recorder Switch   79
        Improving The Sensitivity of a Scanner   79
        Other Analog Modifications for Scanners   80
    Preparation for Scanner Modifications   81
        Qualifications & Experience   81
        Where to Obtain Service Manuals   81
        What the Author Will do if You Run Into Trouble   82
        Soldering Equipment & Skill   83
    Specific Modifications:
        MOD-1 Restoring Cellular: PRO-2004/2005   85
        MOD-2 Speeding Search/Scan Rates: PRO-2004/2005   88
        MOD-3 Another PRO-2004/2005 Speedup Idea   89
        MOD-4 Improving Squelch Action: PRO-2004/2005   91
        MOD-5 Better TAPE REC Quality: PRO-2004/2005   93
        MOD-6 Auto TAPE REC Switch: PRO-2004/2005, etc.   96
        MOD-7 Improved Visibility Keyboard for PRO-2004   102
        MOD-8 Improved Headphone Audio: PRO-2004/2005, Etc.   103
        MOD-9 Disabling the "Beep" in PRO-2004/2005   104
        MOD-10 Making Any Base Scanner More Transportable   106
        MOD-11 Voltage Surge/Spike Protection: All Scanners   107
        MOD-12 An S-Meter for PRO-2004/2005: Method #1   112
        MOD-13 An S-Meter for PRO-2004/2005: Method #2   114
        MOD-14 Interfacing Scanners With Shortwave Receivers   117
        MOD-15 100 Extra Programming Channels for PRO-2004   123
        MOD-16 6,400 Programmable Channels for PRO-2004/2005 &
            Other Scanners, Too   124
        MOD-17 CMT & Other Channels for PRO-2021 (Maybe!)   143
        MOD-18 Restoring Cellular: Realistic PRO-34   145
        MOD-19 3,200 Channels for The Realistic PRO-34   148

MOD-20 Restoring Cellular: Uniden Bearcat BC-950-XLT   152
MOD-21 Restoring Cellular: Uniden Bearcat 200/205XLT   152
MOD-22 Restoring Cellular: Uniden Bearcat BC-760XLT   153
MOD-23 Automated Search/Store: PRO-2004/2005   154
MOD-24 What's in Store for the PRO-2004/2005?   155
   What Else Can They Add to The Unit?
   Wish List for Future Modifications
MOD-25 If You Can't do the Work Yourself or Can't
   Find the Necessary Parts-- All is not Lost   158

Forward by Tom Kneitel

An amazing thing happened recently. Scanner design and technology got to the point where an intrepid group of "scanner hackers" were able to open up their scanners, poke around at their innards, and figure out how to get the equipment to do all sorts of clever things never included by the manufacturers.

This could be something as basic as the cutting out or adding in of a single 25¢ diode at a strategic point in the circuit, or as ambitious as removing one of the scanner's "chips" and replacing it with another that would enable the set to scan 6,400 memory channels. The scanner that opened the door to such techniques was the Realistic PRO-2004, and its later version, the PRO-2005. Because of the popularity and potentials of these two scanners, most of the modification techniques have thusfar been developed in connection with these units. However, a number of mods have also been worked out for the Realistic PRO-34 handheld as well as an array of current Uniden Bearcat scanners.

With these mods you can get the scanners to receive blocked out frequency bands, scan at faster speeds, scan thorough more memory channels, interface with a communications receiver, provide better headphone and tape-out audio quality, and all sorts of other useful things.

Bill Cheek, who many in the communications hobby may also know under his nickname, "Doctor Rigormortis," has done a fine job in developing, collecting, improving, and testing numerous scanner hacker techniques in his laboratory. He then put them into a fine step-by-step format so that the benefits derived from these modifications could be accomplished by scanner owners having little more than basic tools and circuit experience.

Tom Kneitel, K2AES
Editor,
Popular Communications Magazine

Chapter 1

Scanning & The Law

If you'll excuse me, I feel the need to explain a few things to you about the legal ramifications of turbo-supercharging and operating your scanner. Some communications books seem to wander off the track while attempting to do this. Warnings, caveats, and legal mumbo-jumbo have become a way of life in our society, it seems, and sometimes with good reason. In my own case, there's a very, very, good reason. You see, from 1983 to 1988 I published a CB radio magazine called the **Eleven Meter Times & Journal**. During that era, I believed rather strongly in Freedom of the Press.

EMTJ published lots of good radio lore and "insider information" for dedicated and hardcore 27 MHz enthusiasts. The bill of fare was mostly technical stuff worded in ordinary language, and aimed at serious DX chasers who had very little interest in belonging to local CB clubs, monitoring Channel 9, talking to truckers on Channel 19, and the rest of the general run-of-the-mill CB and "goodbuddy" aspects of 27 MHz. The tone of EMTJ was somewhat irreverent, it criticized the "goodbuddy" operators and some of the FCC's methods and tactics in regard to 27 MHz operators. In fact, EMTJ went so far as to say, on occasion, that the FCC meddled with people's civil rights. Then the FCC meddled with mine. Freedom of the Press lost a lot of its meaning for me at that point.

One of the things that the FCC focused in on was information in EMTJ that they claimed could have been put to illegal use by the readers. To paraphrase what one FCC official told me at one point during the proceedings, "We're afraid that people will think it's legal for them to do things written about in your newspaper." It seemed to me that the FCC was looking for some way to make me responsible for people violating the CB Rules!

At another time, another FCC official read off a litany of things he had against me. Then, quite off the cuff, he casually tossed out the thought that, "if the EMTJ somehow just stopped publishing, things could go a lot better for you." Not having the time, resources, nor the inclination to extend this situation any further than necessary (since I had already put two years into it), I took his suggestion with some seriousness. Then, more as an experiment than anything else, I suspended publication of EMTJ and waited to see what would happen next. Several months passed with no further hassles from the FCC. It became apparent that I had made made the right decision, so publication of EMTJ was terminated for good. I haven't heard from them since. The effect of my decision was almost magical, if you know what I mean.

With that most painful experience under my belt, please bear with me as I take a few pages here to deal with some of the legal aspects of scanning, like several areas where people could potentially get into trouble with the units. That way, you won't do anything wrong, and nobody can say that I said it was OK to do this or that. I'm not telling you it's OK to do anything, but I will tell you what I think is illegal in certain areas. I'm not an attorney, and this isn't legal advice, merely my own interpretation of how several laws relate to owning a scanner. Naturally, I'd suggest that you obtain a copy of the laws and read them in their entirety, since you might come up with some additional thoughts and interpretations of your own.

Communications Act of 1934

Let's begin with the Communications Act of 1934, which clearly says certain things about listening to or monitoring radio communications. This law is more than fifty-five years old and it has passed the tests of time at various levels of the federal court system. So, things are pretty well etched into stone and honored by both time and tradition.

1. It's against the law to divulge anything you hear from a radio transmission to anyone else who is not a party to the broadcast. This doesn't really relate to ham, CB, AM/FM/TV broadcasts, or to distress signals. Obviously, any transmission that is intended for general public reception would be exempted from such a restriction.

On the other hand, if you were monitoring a police frequency and heard that a bank robbery was in progress, it would not be legal for you to run next door and tell your neighbor. If you heard that the police just stopped a car that turned out to be stolen, and you recognized the car as the one that belongs to your neighbor, it would not be legal for you to run next door and say that the car was recovered. If your neighbor was sitting in your house listening to your scanner with you when either of these two calls came in, nothing would have to be divulged in order for you to discuss them with one another.

This law has been effectively applied to the dismay of several violators. In one instance, a scanner owner was tuned in on a federal agency frequency and was listening to a surveillance in progress. From the conversations monitored, the scanner owner was able to learn the identity of the suspect. He then called the suspect and tipped him about being under surveillance. This caused major damage to the investigation. However, the federal agents probably had their suspect's phone tapped, for they knew he'd been tipped, and they were able to trace the number that had called in the tip. The scanner owner was tried and convicted of a violation of the Communications Act.

However, it's also interesting to note that many newsrooms as well as the vehicles belonging to radio/TV stations and newspapers are equipped with scanners to monitor public safety and other law enforcement communications. When there's a major crime or fire, sometimes the news vans arrive ahead of the police and fire personnel. Somehow, these people don't seem to get hassled for their monitoring activities, but I'm not a lawyer and I can't give you a list of reasons why.

Ads appear all of the time in hobby publications where scanner owners offer to swap tape recordings of fire or police communications. So far as I know, nobody has ever been prosecuted for this innocent type of monitoring and divulging, although it certainly would appear to actually violate the law. Be careful.

2. It's against the law to make use for personal gain of anything you might overhear in a transmission not addressed to you.

Again, ham and AM/FM/TV broadcasts wouldn't be included in this. And who would fault you for running to the bank to close out your account when the evening news on all three TV networks report the president of your local bank took off to Rio with his secretary and 27 steamer trunks?

Still, you'd clearly run afoul of this law if you acted upon information gleaned from your monitoring of a two-way ship/shore phone call between stockbroker and client. Another violation would be if you overheard on your scanner that the police were going to show up to arrest you, and you then eluded them based upon that information.

If you overheard a cellular phone conversation between a land developer and his partner about a shopping mall they are planning for a certain corner, and you then run to the owner of the land and obtain a five year option to buy it, you've probably

The author evaluating a modification on a Realistic PRO-2004 scanner. New modifications for popular current models are always being devised and tested.

broken at least two laws. The one we're discussing, also one about listening to cellular conversations. We'll get to that one in a little while.

Best bet to be safe and legal here is not to make personal use of anything you might overhear via radio transmissions not intended for public reception.

3. It's against the law to make use for illegal purposes or use in assisting in the commission of a crime, anything you overhear in a radio transmission.

This is one of those deals where if you do one thing, the authorities can bust you for it and then toss in fourteen other things to the list of charges. So, if you use information gleaned from the scanner to aid you in embarking on a life of crime, you should be well aware of the fact that it's a separate violation of the law. So, if you had in mind trying to blackmail the guy down the block because you always overhear his racy carphone chats with his wife's best friend-- like, forget it!

4. The above are the main prohibitions for receive-only situations that seem to be covered by the Communications Act. But, also remember that there are many state and local laws and ordinances that can easily make an instant lawbreaker out of an otherwise innocent scanner hobbyist. For instance, some states prohibit the installation of scanners in vehicles, although there are many variations to the theme. In most cases, hams are excused from the prohibition, and sometimes handheld scanners are allowed as opposed to scanners that are physically "installed" in the vehicles. While all of these restrictive laws plainly stink and reek of being unconstitutional, they are on the books and can easily cost you a fine (or worse) if you get caught.

Other local ordinances may be directed at your base station antenna and the location or height of its installation. Most of the time, federal laws overrule state and local zoning restrictions regarding radio, including antennas, but authorities have figured out all sorts of devilish ways of making their point. I know of one instance of a CB'er whose signals were coming out of a neighbor's stereo. Local authorities decided that they could knock him off their air with a fine and a jail term by determining that his CB operation constituted a "Disturbance of the Peace."

In the case of antennas, while it may be against federal regulations for local authorities to dictate where you can and can't put an antenna, you can be sure that they know how to cook up a list of building and zoning law violations so fast that you'll be content if they agree to let you use an indoor rubber duckie. Neighborhood "protective covenants" can also be used to legislate your antenna into history.

So, I suggest that you obey local laws, even though you may consider them to be stupid or unconstitutional, unless you've got a great attorney, a lot of time and money, and hope to see the story of your plight appearing on page 3 of your hometown newspaper. You may eventually succeed in getting your case heard in federal court, but the sanctions and martyrdom you will experience in doing do will hardly make any victory you might eventually experience seem worthwhile.

Electronic Communications Privacy Act of 1986

And then there's the Electronic Communications Privacy Act of 1986 (ECPA), which covers a range of things you can't now legally monitor. The ECPA expressly forbids monitoring cellular telephone conversations, but that's not all. The ECPA makes it illegal to intercept and monitor any radio frequency communications that are originated from or terminated in a normally private landline. This includes most microwave and satellite transmissions , even broadcast studio-transmitter links in the 26 MHz and 945 MHz bands. It's also illegal now to descramble a deliberately scrambled transmission, even if the common speech-inversion technique is used.

There are exceptions to the ECPA, such as 49 MHz cordless telephone handset signals, CB, ham radio, signals from ships and aircraft, and governmental communications, for instance. Absurd and virtually unenforceable as the ECPA is, many people seem to think that its even unconstitutional. The person to find that out, however, will have to be very rich or very convincing-- or both. Rest assured that someday, someone will eventually get nailed for violating the ECPA, be it an espionage agent or just some average citizen with a scanner who displayed his monitoring prowess on the cellular frequencies to a curious FBI agent.

I am adamantly against the ECPA's provisions that claim the authority to make it illegal for me to monitor the public's airwaves in my own home. My philosophy is simple and is shared by thousands of others. When some person or business beams a radio wave across my property, into my home and through the pores of my skin, then I have the inalienable right to detect that radio wave for personal security reasons. Radio waves, including the basic RF energy and the information carried upon those waves, can be harmful to life and limb. It ought to be my exclusive right to determine the content, frequency, strength, polarity, mode, duration, and other characteristics of that radio wave. For Pete's sake, what if telco wants to run your neighbor's landlines through your living room, across your sofa, and out your bedroom? Radio waves do that every second of the day, except you can't see them or trip over them in the dark. They're still there, though, and unlike wires that can be seen running across the walls and floors, RF is invisible as it passes through your body.

If privacy of communications is required by those who communicate, let it be carried over landlines, or let it be the responsibility of the communicators to encrypt their radio signals. Cost effective, secure, digital encryption equipment is readily available. It works a lot better at providing privacy than does legislation-- and there's no two ways about that!

In any event, it's difficult, if not altogether impossible, to detect violations of the ECPA (and gather evidence of what persons are monitoring in the privacy of their own homes), it is a law that has little chance of being enforced, except in instances of the most flagrant violations. The entire law, or at least certain unconstitutional and/or unenforceable provisions (such as monitoring and interception from a person's own home or private property) should be stricken.

If two persons are conversing on a a clear (unscrambled, unencrypted) radio circuit, would they reasonably and legitimately expect any more right to privacy than if they were standing on a busy sidewalk yelling their thoughts back and forth to one another,

A corner of the author's work center where scanners are taught to do new tricks their designers never envisioned.

or if they were talking in a crowded bus or restaurant? What's the difference if they are doing it while driving around in their cars having their conversation while bombarding you and the environment with 800 MHz radio waves that carry their words? Same situation with satellite transmissions-- waves from outer space that are lambasting your home, property, body. What inalienable right to privacy does that satellite or its owner legitimately expect? Nevertheless, the ECPA was passed by Congress and signed by the President, so it's the law...

We'll have more to say about the ECPA later on in this book.

A little should be noted about modifying scanning receivers. So far as I am aware, there is no law or regulation that relates to what a private party can or cannot do to his or her own receiving equipment; although there are things you aren't supposed to do to transmitters. Back when I was having FCC problems, I asked the FCC Engineer-in-Charge if it was legal to modify the receiver (only) sections of CB radios. I was stunned when he said that he didn't know for certain and would have to "check with Washington." He did go into some lengthy speculation that how the receiver portions of CB radios had to be "certificated" by the FCC, so any modifications would probably violate the certification. There's little similarity, I believe, between CB radios (which are transceivers) and VHF/UHF scanning receivers. The scanning receiver shouldn't have any problems being modified by its owner. However, be aware that the companies that manufacture the scanners maintain an aloof attitude towards their products being modified internally and generally consider modifications as having been cause to void their warranty on the product.

When the ECPA went into effect, it's wording was sufficiently non-specific to create confusion within the ranks of hobbyists and manufacturers. Some came to the conclusion that it was either unethical, immoral, or just plain illegal to possess or modify equipment so that it would be capable of receiving frequencies on which cellular communications in the 800 MHz band where cellular communications take place.

In early 1989, an independent dealer selling new Realistic scanners was advertising that he would, for a nominal fee, furnish the PRO-2004 scanner with the "cellular frequencies restoration" modification. In time, the dealer got a visit from FBI agents who told him that he could no longer provide this pre-sale modification service because it violated the ECPA. The dealer then discontinued the modification service, but, instead, packed a in a do-it-yourself instruction sheet with each PRO-2004 sold. That seemed to satisfy the FBI and also the complaintant (the cellular telephone industry), and there haven't been any additional hassles reported. So, perhaps it might be interpreted that a business can't do pre-sale cellular modifications-- but none of this is really clear. There are certainly reasons to wonder, since several manufacturers produce scanners that come factory-ready to pick up cellular frequencies without any

A close look at the author's all-band monitoring station. Most of this equipment has been substantially modified.

needed modifications. Meanwhile, some scanners are designed to cover these frequencies, but their ability to do so is blocked at the factory and can't be utilized until it is deliberately restored by the clipping of a diode or a resistor-- modifications that are neither recommended, endorsed, nor suggested by the several manufacturers whose products come through this way.

On the whole, the scanning hobby is really one of the least regulated and least hassled pursuits of them all. It's likely to remain that way for some time to come because it's a leisure pursuit you do mostly within the confines of your own home. A little common sense and prudence concerning the laws that relate to scanner usage will help keep things hassle-free for years to come. Our advice is to obtain copies of the various laws that relate to scanning, read those laws, and obey them.

Chapter 2

Tips, Hints & Kinks

Chapter 2

Tips, Hints & Kinks

This chapter is a roundup of the "right stuff" that it takes to scan with the best of 'em. Included here is a wealth of practical tips, hints, kinks, and useful scanner lore without going too far into technical jargon or theoretical ruminations. If you're looking for deep end stuff, it's not here. But if you want a fast-track path to grabbing the absolute most out of the time and resources available to you for scanning, then stand by, because a number of things you'll want to know about are definitely in here somewhere. If I missed something you wanted to know about, please be sure to let me know so that I can include it in a subsequent edition. Now, let's jump right into the single most important thing about radio-- any radio!

Antennas for Scanners

Memorize this: "Give me a $10 radio and a $990 antenna system, and I'll play radio with the best-- and I did." Quote by Yours Truly.

That bit of sage wisdom was my motto in other realms of radio for many, many years. It is most applicable to VHF/UHF scanning, too. By and large, the antenna and coaxial cable make or break the monitoring station. To be very sure, the quality and features of modern scanner radios are worth a lot more than ten bucks, and if all I had to shell out on my scanning hobby was a thousand dollars, well, I'd still have my Realistic PRO-2004 and I'd have to settle for a $580 antenna system. Remember the quotation. When the idea is to successfully copy a single station, the radio drops out of the equation so long as it works, but the antenna and coaxial cable can offer the critical margin of hearing or not hearing.

If you are going to have a decent monitoring station and if you are willing to spend some serious money to do so, then you'll be doing yourself a great disservice if you don't pay attention to the antenna and transmission line. No matter how good (or poor) a receiver you have, the very most it can possibly pick up is only that which enters the jack on the rear of the set. The quality and performance of your antenna system determines as much as of a ratio as 30 to 1 of the signal that enters the receiver! In other words, a superior antenna can deliver thirty times more signal than an inferior one. This is exemplified when you consider that the signal captured by the antenna must be fed down 20, 50, or even 100 feet of coaxial cable. Coaxial cable can do absolutely nothing whatsoever of benefit to the signal, but it can starve your scanner of signal through losses. Even the connectors that mate the antenna to the coax, and the coax to the scanner, can steal valuable signal.

But first, let's talk about antennas. Basically, there are only a few considerations and choices. The very first, and one of the most important, is **height**! There is no substitute for antenna height in terms of low cost versus performance. Every ten feet of increased height makes for a significant improvement in signals, especially those coming from a distance. The more distant a transmitter is, the more improvement in reception you can achieve by simply raising your antenna a few more feet. A cheap antenna can outperform one costing twice as much but mounted ten feet lower.

The rule to keep in mind is get the best antenna your budget will allow and install it as high as it is **safe** and within your budget to do so. I mention safety because it's vitally important that you plan your antenna installation where there is no possibility of it toppling over and coming into contact with electrical wires, or where it could cause damage to life or property in the event it fell during or after its installation. Hams, CB'ers, and scanner owners have been electrocuted when the antenna they were erecting came into contact with a power line while they were holding on. If there is any possibility that any part of your antenna system (including its mounting mast) could make contact at any time with power lines, then find another location where that serious hazard to life does not exist. Also note that all antenna installations should be adequately protected against lightning strikes-- several manufacturers offer good lightning protection devices at reasonable prices for antenna systems.

Height Advantage of Antennas

There's absolutely no substitute for height of antennas in terms of performance gain. The range of VHF/UHF communications over the surface of a (hypothetical) smooth curved earth is essentially line-of-sight, and can be approximated by this equation:

$$D = \sqrt{2 H_T} + \sqrt{2 H_R}$$

where: D is distance in miles
H_T is height of transmitter antenna in feet
H_R is height of receiver antenna in feet

As an example, let's assume a transmitter antenna height of 50 feet and a receiver antenna height of 18 feet. Then we have:

$$D = \sqrt{2(50)} + \sqrt{2(18)}$$
$$= \sqrt{100} + \sqrt{36}$$
$$= 10 + 6$$
$$= 16 \text{ miles}$$

Other variables also enter into the specific equation, such as the gain of both transmit and receive antennas, transmitter output power, atmospheric diffraction, local terrain and geographic features and site losses at each station. Nevertheless, it can be easily seen that as the height of either antenna is increased, the range of communications is also increased.

Rule: Install your monitoring antenna as high as you can safely and economically afford.

Gain of Antennas

Note: For the sake of employing the terminology most commonly encountered within the field of communications, some of the terminology used in this discussion appears to apply to transmitter antennas. In fact, it does, but receiving antennas possess the exact same characteristics. There isn't any difference between the two.

The **gain** rating of an antenna is its power multiplication factor. Some antennas such as discones have no gain (or "0 dB gain"), which is a multiplier of "1." Others, such as log periodic directional antennas, have up to 12 decibel (dB) gain. Yagi type directional antennas can have even more gain. Every 3 dB of gain is a multiplication factor of two. So an antenna with 12 dB gain has a multiplication factor of 16 times that which goes into the antenna. Signals would effectively be multiplied by 16. Inferior or defective antennas can have a negative gain, which, in effect, dvides the input signal (instead of multiplying it) for a loss. Refer to Table 2-1 for an idea of the multiplication factor for various dB gain figures:

Table 2-1

dB GAIN & MULTIPLICATION FACTOR

dB Gain	=	Mult. Factor	dB Gain	=	Mult. Factor
0	=	1.0	7	=	5.0
1	=	1.3	8	=	6.3
2	=	1.6	9	=	8.0
3	=	2.0	10	=	10.0
4	=	2.5	12	=	16.0
5	=	3.2	15	=	32.0
6	=	4.0	20	=	100.0

Bandwidth of Antennas

The **bandwidth** of an antenna is a measure of the frequency spectrum over which the antenna will perform within stated or required limits. Bandwidth and gain don't mix well in the sense that high gain antennas tend to be designed to operate within a narrow band of frequencies while wideband antennas tend to have low gain. Scanner users tend to be interested in having the ability to tune frequencies between 30 MHz and 900 MHz or above, but they're typically limited to using only a single antenna. Therefore, gain must be sacrificed for bandwith. The discone antenna is widebanded with low gain, and is very well suited to the needs of most general interest scanner users.

Selecting a Scanner Antenna

So how do you know which is a good antenna and which isn't? Scanner users have less to worry about here than most other communications hobbysts. With CB'ers and hams, there is a great deal of performance difference from one antenna model to another. There are so many models to choose from, and some seem to be pure junk despite the glowing terms in which they are described by their producers. Scanner users have a smaller selection from which to pick. Also, there is the very nature of VHF/UHF monitoring which normally doesn't include long distances.

For example, at my own monitoring station in San Diego, I can pick up most of the public service stations in Los Angeles, about 140 miles to the north, but I rarely bother to listen. The local San Diego scene is my interest, not some alien and unfamiliar city that's about a three hour drive away. So, I don't require a super-duper antenna offering lots of gain. You may have a similar situation, finding that (thanks to repeaters and high powered transmitters) you can pick up almost everything you want to hear in your area using no more of an antenna than a telescoping whip attached to the back of your scanner. On the other hand, that antenna isn't going to pick up non-repeated signals from any distance. So if all you're interested in is the local 162 MHz NOAA weather forecast, use the indoor antenna that came with your scanner. You'll hear quite a bit, and it might be enough to satisfy your needs.

If you want to hear on-scene disaster and emergency communications, mobile units, handhelds, adjacent cities and counties, and other comms over a wide area then you'll have to think "antenna," and that's all there is to it. So, even though I rarely bother to listen to comms in Los Angeles, I still wish to have the capabilities to do so should I wish to for a specific purpose. Southern California has no shortage of earthquakes, forest fires, and other assorted calamities that motivates me to want to be able to hear what's going on outside of my immediate area. When something is going on that has even the remotest possibility of negatively impacting on the environment, or on my family's pursuit of life, liberty, and happiness, then I want to hear for myself exactly what is going on. It's not going to be the scanner that makes the difference, it's going to be the antenna.

Types of Antennas

Antennas come in two general types: onmidirectional and directional. **Omni** comes from the Greek, meaning "all," so an omnidirectional antenna receives (or transmits) equally well in all directions. An analogy is that a bare light bulb is an omnidirectional light source. Sometimes omnidirectional antennas are miscalled "nondirectional" antennas, which seems to me a strange contradiction in terms describing an antenna that won't work in any direction.

Directional antennas, by definition and design, favor some directions more than others. To continue the analogy, bring a mirror up to one side of a bare light bulb. Now it will be rather dark behind the mirror while the area across from the front of the mirror will be almost twice as illuminated as without the mirror. That increase in illumination is called **gain**. Gain comes by concentrating light (or radio energy) in some directions to the exclusion of others.

Omnidirectional antennas have little or no gain, they offer uniform performance pretty much in a 360° circle. Omni antennas can have some gain, depending on the use of designs that can decrease radiation in angles straight up and straight down so as to compress the radiation pattern into a "doughnut" shape, with most of the signal pattern concentrated equally around in a 360° circle.

Directional antennas have gain or enhanced performance in the direction where they are pointed. They have diminished performance in other directions. Directional antennas are therefore usually mounted on rotating devices that permit the operator to aim the antenna at different areas, and change the direction at will. Without a rotator, a directional antenna offers permanent coverage only in the one direction you have it installed, unless you can lean out of your radio room window and torque the mast around by hand (a method radio operators call an Armstrong rotator). However, if you live out in the boonies, and the only scanner action of interest to you comes from a town located off in the distance to the east of you, you may well find that a directional antenna left pointed eastward is just what you've always wanted.

Since directional antennas decrease energy in some directions and increase it in others, there is another specification that is somewhat related to gain, but deserves separate mention. This is **Front-to-Back** (F/B) ratio and is a comparison of a signal measurement on the front side of the antenna, to another measurement of the same signal, but made when the signal is pointed in the opposite direction. If there is ten times more signal on the antenna's front than on the rear, then it is said to have an F/B ratio of "10." Hobbyists sometimes sometimes refer to the F/B ratio as "rejection," which is a rather good term.

Gain and F/B ratio are about the primary specs of directional antennas if you ever consider getting one. Both specs are normally stated in decibels (dB), and usually the larger the number, the better. Unfortunately, scanner antennas are often offered for sale in a manner similar to TV antennas in that the manufacturers aren't always inclined to quote performance specs in a way that you can easily know what you're getting, nor compare one antenna to the next. Sometimes getting gain and F/B ratios for directional antennas is like pulling teeth.

Most hobbyists seem to think that gain is the big deal about directional antennas, and appear to be little interested in the F/B ratio even though it may well be more worthy of consideration than the gain figure. For example, if you live near a VHF or UHF repeater site or a high powered transmitter, your receiver could be plagued with intermod and strong signal overload (interference) which would "desensitize" your receiver and impede hearing distant signals. A directional antenna can be positioned with its rear to the source of interference or strong signal to reject enough of the unwanted signal for your scanner to perform normally in other directions.

When shopping for a scanner antenna, you need common sense and a little knowledge. Here is some antenna lore and hard, cold, brass tacks: Omnidirectional antennas fall into three main categories--

A. Groundplane, distinguished by one or more parallel, vertical elements of varying lengths, plus three or four horizontal or drooping radial elements spaced $120°$ or $90°$ apart, respectively. Coaxial cable feed is normally at the bottom of the antenna, just under the horizontal radials. Generally, the groundplane has the highest gain of omni antennas at up to +3 dB. Certain designs can approach 6 dB gain, but most omni antennas marketed for scanner users have 3 dB or less. Bandwidth of a typical scanner groundplane (such as Radio Shack 20-176) is relatively narrow, but with the addition of extra (two or more) vertical elements (such as the Radio Shack 20-014), bandwidth can be much wider.

B. Dipole is a single element antenna separated in (or near) the center by an insulator. Coaxial cable feed is at the insulator. The gain of a dipole is low (+2 dB), and the bandwidth is relatively narrow compared to groundplanes and discones. The bandwidth of a dipole can be expanded with a little loss of gain if the feedpoint is moved off-center. An example of an offset dipole is the Grove "Omni" (Model ANT-5). Later in this chapter are some home construction projects for off-center dipoles.

C. Discone, which is characterized by four to eight horizontal radials at the top with an equal number of elements connected below the horizontal section which droop or slope downward and outward at an angle. A discone is the widest banded of the omni class of antennas, but also the lowest in gain (0 dB). A discone sometimes has a single vertical element mounted above the horizontal radials, which increases VHF "low band" (30 to 50 MHz) performance without sacrificing performance on the higher bands. An example of the latter is the ICOM AH7000. Radio Shack's 20-013 is a discone, but without the vertical element. Later in this chapter is a modification to expand the bandwidth of the Radio Shack discone. Another popular discone having a vertical element is the Palomar D-130. Discones for scanners are typically rated to perform from 25 MHz to 1300 MHz.

Directional antennas for VHF/UHF are generally confined to two categories:

A. Yagi-Uda, usually called simply a "yagi," is characterized by a single, horizontal "boom" used as the mounting structure to support three or more vertical elements perpindicular to the boom and spaced at varying (but critical) distances apart. Yagi antennas offer high gain and good F/B ratio, depending upon the number of elements and overall design. Without special design techniques, yagi antennas are relatively narrow banded. Many companies make yagi antennas for the commercial VHF and UHF markets and these can be pressed into service for scanners, if narrow band operation is satisfactory.

One of several commercially available discones available to the scannist for base station use. These are omni-directional types.

B. Log-periodic antennas, at first glance, look similar to yagi types. However, they have many more vertical elements than yagis, and the elements are equally spaced apart, with the ones at the front being radically shorter than those in the middle and the other end. Yagi antennas have different spacings between the elements, the lengths of which vary very slightly from one element to the other. Unlike with the log-periodic, a yagi's elements-- when seen from a distance-- appear to be about the same length. Log periodics have lower gain and lower F/B ratio than yagis, but much wider bandwidth. Overall performance is proportional to the number of elements. Scanner users don't have much to select from in this category of antenna, but Orion Hi-Tech and Hy-Gain/Telex distribute several models.

Still, the discone, remains the best overall antenna for the average to serious scanner owner if only a single antenna is possible or desired. Ideally, you'd want to use a separate antenna for each band of frequencies to be monitored-- but owners of wideband receivers would be hard-pressed to erect five to ten special antennas, with the necessary switching arrangements that would be required. Therefore, the discone comes highly recommended, and it offers the plus of presenting a flat 50-ohm load to the scanner across its entire bandwidth. This is a strong plus since other antennas touted as being "wide band" are often notorious for variable and uncertain impedance, depending upon the band and the antenna design. This can be the source of major performance loss for critical situations.

TV Antennas for Scanners

If you have the space and the budget, a directional antenna would be a handy addition to your monitoring post for some situations. If you want to try a directional antenna at a modest cost without getting in over your head, your best bet might be a TV/FM antenna! Yes! Look at it this way, VHF TV antennas are designed for FM broadcast plus TV Channels 2 through 13 which are:

TV Channels 2 to 6 = 54 to 88 MHz
FM Broadcast = 88 to 108 MHz
TV Channels 7 to 13 = 174 to 216 MHz

TV antennas with UHF capability in addition to VHF include:
TV Channels 14 to 83 = 470 to 890 MHz

Therefore, an ordinary combo VHF/UHF TV plus FM antenna could be a perfect second antenna-- with a slight hitch. In order to receive the horizontally polarized TV and FM broadcast signals, TV antennas are designed to be mounted with their antenna elements in a horizontal plane, that is, parallel with the ground. VHF/UHF communications signals, however, utilize vertical polarization. That means the TV antenna must be rotated 90° to be correctly positioned as a scanner directional beam antenna. This is easy to do-- Install a 2 to 3 foot "mast" section in the TV antenna mounting bracket. Turn this mast so that it is parallel with the ground (and the TV elements perpindicular to the ground) and bolt the short mast to a normal, vertical mast section or the tip of your supporting structure, such as push-up/telescoping mast, or whatever). The bolt can be a simple U-bolt.

An ideal "starter" TV antenna for your scanner could include Radio Shack 15-1641,

15-1642, or 15-1643. Of course, you could go on up the list with any of their several high performance, deep-fringe models, too.

There are three important considerations if you use a TV/FM antnna as a scanner beam:

A. Install the TV antenna in a vertical polarity as described, at least 2-feet away (off to the side of) from the main mast or antenna support.

B. Install a matching transformer (balun) right at the terminals of the TV antenna. This is an adaptor to match the terminals of the antenna to a coaxial cable downfeed. The balun has a short pigtail of "flat lead" on one end and a female Type F connector on the other end. Use Radio Shack's 15-1140, 15-1143, or equivalent.

C. Use good quality coaxial cable and connectors in your downfeed. One very useful setup could include 50-feet of preassembled low-loss coax with Type F connectors (Radio Shack 15-1526, for example, or equivalent). One end of this coax will connect directly to the antenna at the balun matching transformer mentioned in Step B. Run the coax down to your station and use a Radio Shack 278-251 (BNC to Female F Adapter) on the scanner end of the coax. The adapter then easily connects to the BNC connector used on the back of just about all currently made scanners.

Note: Even if your scanner doesn't use a BNC type connector, Radio Shack has a wide selection of adapters that will match the Type F connector on the coaxial cable to most types of antenna jacks found on scanners.

Naturally, you'll want to be able to point your antenna around a 360° azimuth, so put a rotator on your shopping list. Radio Shack's 15-1225 should handle just about anything but the very largest monster "scanner beams." Here's a hint when you've got a rotatable beam-- when they expect high winds, point the beam into the direction of the wind rather than permit the wind to hit the antenna broadside. This presents the wind with the smallest possible surface area upon which to blow, and it could well save your antenna from being damaged. Of course, if you live in a rural area and most of what you want to hear in one general direction from your station, then forget about the rotator. Just install the antenna so that it's aimed at the action.

Installing Antennas

Scanner antennas are usually light and fairly compact, so a good support is the common telescoping mast that TV antennas are mounted on. As towers and masts go, it's a rather inexpensive way to go, and it's easy to work with. A major disadvantage is that it needs guy wires at the joint end of each section. Putting up fifty feet of telescoping mast is going to take the best efforts of several agile people and is hardly to be considered something you're going to do on a Saturday morning. A twenty-footer, however, is a snap. Your job is to determine what you can afford, what you know how to do, and what you have to work with. The key is to maximize.

For instance, I've seen some 50-foot telescoping masts installed at ground level, which is easier than putting one on the roof. But, in some of these instances, a 20-foot mast installed on the highest part of the house would have been higher and no more complicated. Sometimes biggest doesn't always mean best. Install the best antenna your budget will permit, and do it as high as is safe and reasonable.

Constructing/Modifying Antennas

A. Radio Shack Discone (#20-013). The useful performance bandwidth of this antenna can be extended down to 25 MHz with a simple modification.

First, you must understand that the published specs for this antenna indicate a bandwidth covering 25 to 1300 MHz, but references and performance curves are not available so the claim is hard to refute. My own extenstive tests and evaluations of this discone suggest that performance falls off rather sharply below 100 MHz. This is understandable when the Radio Shack discone is compared to others (such as the ICOM

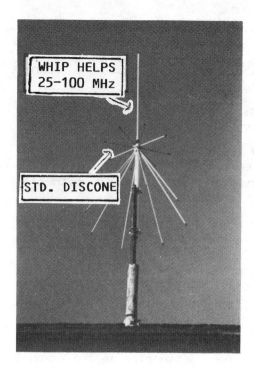

A discone antenna is ideal for monitoring the 25 to 1300 MHz area of the spectrum, however unless the antenna sports a vertical element at its top, coverage below 100 MHz may suffer. In fact, you might even be able to add this vertical element to some discones that didn't have it as part of their original design. See the text for more information.

AH7000) that offer similar bandwidth specs. If you look at the ICOM unit, however, you'll see a typical discone shape, but also a single, base-loaded vertical element that rises above the eight horizontal top radials. It is this extra vertical element that contributes to performance below 50 MHz. The Radio Shack discone doesn't have this element, so reception in the 30 to 50 MHz scanner band suffers.

This shortcoming was discovered when my Realistic PRO-2004 and discone combo couldn't do very well on 27 MHz CB, the 28 MHz ham band, or the 30 to 50 MHz scanner band. All sorts of tests and checks on the antenna were run, but no malfunction was found. Then I just happened to notice how a couple of other discone models had that extra vertical element. A probable solution loomed on the horizon.

The remedy wasn't very difficult and there may well be several other ways to approach a solution. At the top of the Radio Shack discone (just above the eight horizontal radials) is a black plastic cap that I removed to reveal a male 3/8" threaded stub! Will wonders never cease? A wonderful place to mount a vertical element of some kind.

An el-cheapo, no-brand, base-loaded CB magnet-mount whip antenna was salvaged for the job. The base-load coil assembly was unscrewed, retaining the stainless steel whip and base-load coil as a unit. The mag mount base and coax assembly was discarded. There was a female threaded unit recessed up inside the base-coil and it appeared to be a match for the threaded stub at the top of the discone. The threaded stub on the discone was only about ½" long and didn't protrude far enough into the bottom of the base load coil to mate with the female threads. That problem was easily solved when I threaded a short female 3/8" union on to the stub at the top of the discone. Into the union went a short piece of male 3/8" threaded stock. Presto! The CB base-load antenna neatly threaded down for a perfect fit atop the Radio Shack discone.

The modified joint was then weatherproofed, the antenna was reinstalled, and another series of tests conducted. Sure enough, low band VHF, CB, 28 MHz ham signals were dramatically improved (over 9 dB) all the way down to 25 MHz! "Before and after" comparisons indicated that performance above 50 MHz didn't suffer because of the mofification.

An assortment of antennas for handheld scanners. The stock rubberized ones that come with some scanners don't always do a great job. Telescoping metal whips work better but are prone to getting damaged with normal handheld use. High performance rubberized replacement whips are a good compromise for use in the field.

As mentioned previously, there could be other ways of accomplishing the same thing with good results. The key is to make use of the male 3/8" threaded stub that sticks out of the top of the Radio Shack discone. Some kind of coil, CB base load, or whatever you can manufacture, should be attached to this stub. A whip not more than 3 to 4 feet long should be attached to the top side of the coil. Experiment with both the coil and the length of the whip to maximize the performance of the discone for the middle of your favorite VHF low-band-- or for about 40 MHz if you don't have a special favorite. The number of turns and the spacing between the turns of coil can be varied for the sake of experimentation.

B. Rubber Duck   antennas supplied with handheld scanners are sometimes those 6 or 7 inch types which are rather poor performers. With its flexibility and durability, the rubberized antenna is probably your best bet if you are "on the move" with your handheld scanner. If you like the convenience of a rubberized antenna, you can probably improve reception by replacing your stock 6 or 7 inch rubber duck with a longer, helical, top-loaded high-efficiency unit such as the CRB Scantenna.

If your handheld scanner is in use, but not on the move, you can try a semi-rigid, telescoping rod type antenna. This type of antenna probably isn't going to stand up well when you're on the go with your handheld scanner, but will provide good reception and service when your scanner is in use on your desk at the home base. Radio Shack offers two telescoping rod whips with BNC connectors, 20-006 and 20-008. Grove offers their ANT-8B. All are good performers.

Experimenters might want to build their own. You need a male BNC connector and a replacement type, telescoping rod antenna, a little resourcefulness and ingenuity. I've made a dozen or so now with great success. The secret is to use "5-minute epoxy" in the base of the BNC connector after the rod antenna has been soldered to the pin of the BNC connector. The epoxy insulates the whip from the ground shell of the BNC connector and also provides strength for the base of the antenna. After it has cured, the epoxy base can be sanded and polished for appearances.

Most any telescoping rod antenna (such as Radio Shack 270-1401) can be used so long as you can figure out precisely how to connect the base of the rod antenna to the center pin of the BNC connector. Generally speaking, the longer the rod antenna, the better. Also, the more sections in the rod, the better. That way, when not in use, the antenna may be collapsed down into a short length. An extended length of 30 to 40 inches is about right, and the number of telescoping sections should be five or more.

C. All-Occasion, Portable Base Station Antenna. Here's an effective scanner antenna (off center-fed dipole) that can be easily fabricated for a few bucks. It's an

The Grove "Omni" antenna is an off-center fed dipole. Feeding a dipole off-center adds some extra features to the trusty old dipole design. See the text for furter information.

excellent antenna for apartment dwellers, for temporary field use or for emergencies. It can be rolled-up and tucked away as a spare or backup for a time of need.

1. Get a 300 ohm-to-75 ohm matching transformer such as Radio Shack's 15-1140, plus a short length of preassembled RF/video cable such as Radio Shack's 15-1530 (8 ft.); 15-1531 (16 ft.); or 15-1534 (25 ft.).

2. Solder an 18" wire (14 to 22 ga.) to one of the flat-lead terminals of the matching transformer.

3. Solder a 48" wire (14 to 22 ga.) to the other flat-lead terminal of the matching transformer.

4. Connect one end of the preassembled RF/video cable to the screw-on Type F connector side of the matching transformer.

5. Connect an adaptor designed to mate your scanner's antenna jack to the Type F connector at the other end of the preassembled RF/video cable. If your scanner has the currently popular BNC type antenna jack, then you'll want Radio Shack's BNC male-to-female #278-251.

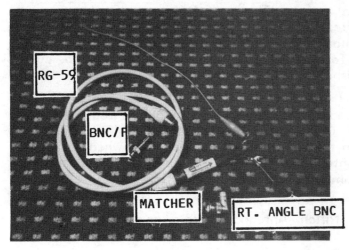

A couple of easily available components can get put together to make an off-center fed dipole. You'll need some RG-59 cable, a BNC-to-F female adaptor, a TV matching transformer, and possibly a right angle BNC adaptor.

6. Connect the adapter-end of the preassembled RF/video cable to the antenna connector on your scanner.

7. String out the RF/video cable and position the two wires of the antenna end on end with respect to each other (a straight line), and then position the entire antenna assembly straight up and down, perpendicular to the ground/sky. The RF/video cable will "feed" the antenna in the middle. The ends of the antenna can be secured to a wall with thumb tacks, or to a tree with whatever's handy-- string, rubber bands, paper clips, nails, wire twisters, etc.

Note: Put the antenna up as high as you possibly can. Either the long wire or the short one can be on top; you can switch them around to see which gives you best results. You can even experiment with different lengths of wire instead of 18" and 48" lengths as specified here. Just don't cut the two wires the same length or even close because then you'll have a narrow-band dipole. Chances are that you'll want this antenna to cover as much of the VHF/UHF spectrum as possible, and radically different lengths for the two wires broaden the bandwidth of the device. When experimenting, use longer lengths for the lower frequencies and shorter lengths for the higher frequencies-- but don't go shorter than 12" or longer than 84" for either wire or you'll run into a point of diminishing returns.

Performance note: Not long ago I built one of these antennas using a 4-ft RF/video cable, BNC/F adaptor, matching transformer, with a 24" alligator clip lead for one "leg" of the antenna, and two 24" clip leads in series for the longer "leg." When connected to a Realistic PRO-34 handheld scanner on my service bench, the 162.55 MHz NOAA weather station in Los Angeles (100+ miles away) came booming in, although it had not been accessable using the whip supplied with the scanner, and only marginally using a telescoping rod type antenna.

D. Another Indoor or Emergency Antenna. If your resources, tools, and time are severely limited, here's another one of those antennas that's quick, will work in a pinch, and does a reasonably good job. All you need is some wire, a piece of RG-58 or RG-59 coaxial cable and the appropriate matching connector to mate with the antenna jack on your scanner. For convenience, the following are the Radio Shack stock numbers for either coax and proper matching BNC connectors:

RG-58 coax cable: 278-1326   BNC male connector: 278-103 or 278-185
RG-59 coax cable: 278-1327   BNC male connector: 278-104

The coax can be any length. If several feet or more, you'll have some ability to vary the height and make position adjustments.

Method 1--

1. Attach a BNC connector to one end of the coax (or other appropriate connector to match the antenna jack on your scanning receiver).

2. Prepare the loose end of the coax by trimming off about 3" of the black outer insulation. Comb or carefully unravel the exposed shielding conductor; separate it from the exposed center conductor and twist it together. Strip about 1" of the white inner insulation from the center conductor.

3. Install an 18" wire (solder it or make a mechanical splice) to the coax shield that you twisted together in Step 2.

4. Install a 48" wire (solder it or make a mechanical splice to the coax center (inner) conductor.

5. String out the coax cable and position the two wires of the antenna end-on-end with respect to one another (a straight line), and then mount the entire antenna assembly in a vertical (up/down) plane. perpendicular to ground/sky. The coax cable will feed the signals to your scanner. The ends of the antenna can be secured to a wall with thumb tacks, or to a tree with whatever's handy.

Note: Height, position, and lengths of the wires may be experimented with for enhanced performance on frequencies of particular interest.

Method 2--

This technique requires only a section of coax (RG-58 or RG-59) at least ten feet long and a BNC or other appropriate connector to match the antenna jack on your scanner. Wire, solder, and splicing are not required.

1. Attach the appropriate BNC connector to one end of the coax (or other connector to match the antenna jack on your scanner).

2. At the free or loose end of the coax, measure back toward the connector exactly 48 inches. Carefully cut or slice the black outer insulation in a circle around the coax at the 48" point-- <u>without nicking or cutting the inner braided shield</u>. Then, slice through the insulation, lengthwise, in a straight line for 48"-- all the way back to the free end, again <u>without nicking or cutting the inner braided shield</u>. Remove and discard the black outer insulation material.

3. Mark a point 30" from the free end of the coax and carefully cut or clip through the outer shield braid until it comes free. DO NOT cut into the inner insulation nor the inner (center) conductor. Remove and discard the 30" loose section of shield braid.

4. Bunch up or loosen the outer shield at the 48" point where the insulation was removed in Step 2. Use a nail or other pointed object to "work" an opening into the shield braid. Don't cut or damage any of the wires in the braid, just separate the braid wire bundles so that a hole forms, and so that you can see the inner insulated wire.

5. Slip the nail or a thin pointed object into the hole and under the inner wire. Using leverage, pry and carefully work the center wire out through the hole formed in the shielding braid. Once you get it extracted a couple of inches, a loop will form which can then be pulled the rest of the way by hand. Pull the 48" center wire out through the hole in the shielding braid, taking care not to damage the shield.

6. Pull (stretch) the shielding braid in one hand against the now free inner wire in the other hand. That's it, you're done. String out the coax cable and position the 18" free braid and the 48" inner conductor end-on-end with respect to each other (a straight line), and then position the entire assembly perpendicular to ground and sky. The ends of the antenna may be secured to a wall with thumb tacks, or to a tree with anything handy. Connect the coax to your scanner and you're set to monitor.

Coaxial Cable (Feedline) & Connectors

Any discussion of antennas calls out for some attention also given to the feedline between the antenna and the scanner. You can have the best antenna available, but if the link that connects it to your scanner isn't any good, then the antenna itself isn't any good. It's as simple as that. You may quote me.

Coaxial cables can be compared to automobiles: there are all kinds, sizes and prices, and generally, you get what you pay for. All feedline, no matter how good, has signal loss. In other words, a certain level of signal is deposited by the antenna into one end of the cable, but by the time it comes out at other end, where the scanner is, the strength of the signal is lower. Good coaxial cable has a lot less of this loss than poor or even "average" cable, you you'll do well to get the best you can afford.

Let's start with what you **don't** want: RG-58. Never, absolutely never, use RG-58 in the normal course of operation as the coax of choice at your monitoring station unless its length is less than 10 to 15 feet. RG-58 has too much loss at VHF/UHF, and there's little sense in losing 75% to 98% of the signal before it even gets to your scanner, and that's exactly what RG-58 cable does, especially at 200 MHz and up!

All coax cables are frequency sensitive; that is, the signal losses increase as the frequency increases. For example, RG-58 cable can be adequate for your 100 kHz to

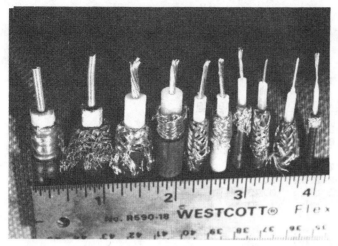
Various 50-ohm coaxial cables don't come at the same price nor offer the same service to the scanner user. At the left here is Heliax hardline-- good stuff but expensive and difficult to work with. Over towards the right are types like RG-58, which are cheap. They also work poorly at VHF and UHF frequencies.

30 MHz communications receiver, or your 27 MHz CB operations, but it just doesn't cut the mustard for scanner use. Write that down.

But, there's a problem for many scanner owners who just don't know where to get tech products such as coax cable except at Radio Shack. While Radio Shack does have an enormous number of products at good prices, and it's convenient to shop there, the serious scannist sometimes finds the need for specialty items that aren't available at Radio Shack. This is especially true when it comes to scanner antennas and coaxial cable. But all is not lost! Bear with me.

Scanning receivers are designed for 50-ohm coax cable and 50-ohm antennas, and we generally try to stick with that standard, but for receive-only applications where loss in the coax is more important than 50 or 75 ohms impedance matching, then Radio Shack does have a cable of interest to you: RG-59 and RG-6. These are 75-ohm types and are normally used with video and television, but they're actually better for scanners than the 50-ohm coax types available at Radio Shack. The lower loss of the RG-59 and RG-6 will more than make up for the slight impedance mismatch caused by the 75 ohm cable. I use the stuff myself in certain instances and it works fine-- runs rings around RG-58 at scanner frequencies.

Radio Shack's RG-8/M and RG-8 can be used for scanners in some instances, but these cables are thick and therefore difficult to attach to BNC connectors, and they do have more loss than RG-59 or RG-6. If you're not a perfectionist, but still wish to minimize signal loss in your coax, then I suggest Radio Shack's preassembled 50-ft. cable as mentioned under the TV antenna discussion. It's good stuff and you can pick up whatever adapters you require to mate this coax to your antenna and scanner.

For the total purists and deadly serious scannists, Radio Shack will have to be forsaken for a trip downtown in a larger city to an electronics distributor who has a coaxial cable selection, or else look through mail order catalogs available from communications equipment dealers. If your interest in VHF/UHF monitoring extends into the 1000 MHz region, then the only way to go in style is with what is called "hardline" or "heliax" cables. Hardline cables are made by a number of companies including Andrew Corporation, Cablewave, and Scientific-Atlanta. The Andrew LDF2-50A is a good, all around hardline, and is roughly the same physical size as RG-8. Also, it is not quite as costly as the larger hardlines.

Hardline coax is difficult to work with, however, because it is relatively rigid and the connectors are very special exotic types that cost a king's ransom. Hardline is expensive, too, starting at $2 per foot and going upward from there. Sometimes hardline and its associated special connectors can be purchased at reasonable prices on the "military surplus" market, but most of the time the cost is beyond the means of the average scanner user.

Table 2-2

A COMPARISON OF COAXIAL CABLES

CABLE MFGR & TYPE	O.D. (Size)	LOSS PER 100-ft (dB)		VELOCITY FACTOR	% SHIELD	RATING
Andrew LDF5-50A (Hardline)	.875	50 MHz: 100 MHz: 400 MHz: 1000 MHz	0.25 0.37 0.75 1.50	89%	100%	Superior
Andrew LDF4-50A (Hardline(	.500	50 MHz: 100 MHz: 400 MHz: 1000 MHz	0.48 0.68 1.50 2.34	88%	100%	Superior
Andrew LDF2-50A (Hardline)	.440	50 MHz: 100 MHz: 400 MHz: 1000 MHz	0.75 1.05 2.16 3.50	88%	100%	Superior
Belden 9913 (RG-8 similar)	.405	50 MHz: 100 MHz: 400 MHz: 1000 MHz	0.90 1.40 2.60 4.50	89%	100%	Excellent
Belden 8214 (RG-8 similar)	.405	50 MHz: 100 MHz: 400 MHz: 1000 MHz:	1.20 1.80 4.20 7.00	78%	97%	Very Good
Radio Shack RG-6	.266	50 MHz: 100 MHz: 500 MHz: 900 MHz:	1.20 1.80 5.10 8.30	75%	?	Good
Radio Shack RG-59	.242	50 MHz: 100 MHz: 500 MHz: 900 MHz:	1.80 2.80 7.50 10.60	75%	?	Fair
Radio Shack RG-8	.405	50 MHz: 100 MHz: 400 MHz: 1000 MHz:	1.70 2.50 6.50 12.00	66%	?	Marginal
Radio Shack RG-8/M	.242	50 MHz: 100 MHz: 400 MHz: 1000 MHz:	2.20 3.00 7.50 13.00	75%	?	Marginal
Radio Shack RG-58	.196	50 MHz: 100 MHz: 400 MHz: 1000 MHz:	4.00 5.30 12.00 22.00	75%	?	Poor

? = Radio Shack does not specify the percent shielding of their coax.

No problem, for there are satisfactory alternatives to hardline that appeal to even some purists. Belden makes a wide assortment of wire and cable, and two varieties are especially of interest for those installing a high-efficiency monitoring station. Refer to Table 2-2 for types and characteristics of a number of cables (including the two from Belden) so you can compare for yourself.

There are four specifications of interest for coax: diameter or size; loss per 100-ft. at various frequencies; velocity factor and percent shielding. The most important specification is **loss**; the least important is velocity factor. Radio waves travel at the speed of light in free space, but they slow down when they have to move along a cable. Velocity factor is the percent of speed of light at which the coax will conduct RF energy. The percent shielding is important, but can be disregarded if **loss** is carefully considered.

Now you need to know the importance of "dB loss." The following table converts "dB loss" into percent loss, which is easier to understand:

Table 2-3

dB LOSS to PERCENT LOSS, CONVERSION

dB Loss	=	Percent Loss	dB Loss	=	Percent Loss
1		21	8		84
2		37	9		87
3		50	10		90
4		60	12		94
5		68	15		97
6		75	20		99
7		80	25		99.7

Don't let this boggle you. In radio, losses are more tolerable than the same percentage taken from your wallet or from your dinner plate. Lose half of your money or half the food on your plate, and you've definitely got a problem. But lose half of your radio signal and the radio can hardly tell the difference! I'm thinking about "3 dB" of loss, which (even though a 50% loss) is relatively insignificant. The radio just barely recognizes it. For example, if a distant transmitter reduces its power by half, your radio will will scarcely be able to tell the difference. The same marginal effect occurs when half the signal is lost in your coax cable-- but you know that at least some of it is going to be lost there anyway, and that's all there is to it.

The idea then is to reduce and minimize that loss. So, to put things into perspective when you look over the coaxial comparison table-- at over 99% signal loss, you can see why you might not want to use RG-58 for 800 MHz and above! Even in the 118 to 137 MHz VHF aircraft band, RG-58 losses will be 75% or more, and that starts to become significant.

Another thing to pay attention to is **length.** Table 2-2 incicates losses in coax cables per 100 feet. Well, 100 feet is a lot of coax, and most scanner installations probably need less than 50 feet of cable. To determine your cable's loss, divide the "dB loss" figure by 100, then multiply by your cable's exact length. That will give you the dB loss for the cable at your own station. Then, refer to Table 2-3 to find your approximate loss in actual percent of signal. If you're losing more than 70% of the signals on a frequency in which you are particularly interested, then you should seriously consider upgrading your coax by one or two levels.

Based upon Table 2-2, which compares coaxial cables, my suggestions for scanner coaxial cables are shown in Table 2-4.

Table 2-4

CHOOSING YOUR COAXIAL CABLE

TYPE OF SCANNIST YOU ARE:	USE THIS TYPE OF COAX CABLE:
Professional "purist":	Andrew LDF4-50A or LDF2-50A
Dedicated hobby "purist":	Belden 9913
A very serious hobbyist:	Belden 8214
An interested hobbyist:	Radio Shack RG-6, RG-59, or RG-8
A casual hobbyist:	Radio Shack RG-6, RG-59, or RG-8/M
Occasional hobbyist:	Back-of-the-set antenna & no coax

Connectors & Splices

Here's something else to write down: NEVER splice your coax! Splices ruin cable and destroy its characteristics, not to mention create intolerable losses. There are two kinds of splices, and whatever results from either one won't be good. One is where you solder the ends together and then wrap tape around the cut. Never do that! If your coaxial cable is too short, don't splice in a little more. Replace the whole thing with a section cut to the proper length.

The second splicing method isn't quite as awful, but it should still be avoided because of the losses that will result. This approach involves joining two sections of coax which already have male connectors installed on the ends. A "barrel" coupler (such as Radio Shack 278-1369) is used to join the two ends together to make one longer section of cable. This might be OK in a pinch for temporary use, but remember that it's still a splice and all splices add their own share of unnecessary extra loss.

All connectors have extra loss, too. You just can't get away without a minimum of two connectors on your coax, one at each end. You're stuck with those, but add more loss from splices, other equipment hooked in-line, etc. and you definitely will pay a price. The total losses can add up to something that would shock you!

Quality and type of connectors also makes a difference in performance. Until a few years ago, most base station scanners were made with so-called "Motorola" jacks for the antenna port. These connectors may be dandy for your car's AM broadcast receiver, but there is a loss in them at VHF/UHF frequencies. Even the popular PL-259/SO-239 connector combination that's used in ham and CB radio has serious limitations at UHF frequencies.

It's no exotic frill or a plot by a radical engineer that caused the ICOM R-7000 to be designed with a Type N connector at the antenna port, and otherwise most current scanners to be produced with BNC connectors. Type N and BNC connectors are well suited to VHF/UHF frequencies because they exhibit lower loss and do not fritter away a little bits of that valuable signal you want to hear! The point is that any mechanical connection between the antenna and the input on the radio will consume some of the signal that passes through it; this in addition to the losses directly associated with the coax. For whatever it's worth, a 81% loss from the coax and another 25% loss from poor or inferior connectors equals a 94% combined loss (.75 + <.25 x .25>). So, when it comes to scanners, avoid Motorola plugs and jacks, and even PL-259/SO-239 connectors. The best connectors for VHF and UHF, in order of preference, are: Type N, Type BNC, Type F. The type of connectors normally used with cellular telephones and their antennas are Type TNC-- a cellular antenna may be used for general scanning the 806 to 912 MHz range if you mate the TNC connector to your scanner's BNC input with a Radio Shack 278-144 adapter.

These days, the scannist may find it useful to become familiar with the various types of coaxial connectors, joiners, and adapters that usually are required when assembling a monitoring station. Terms such as BNC, Type F, Type N, and PL-259 soon become a part of your jargon.

A final note on the subject. Adapters really should be avoided whenever possible. Yes, I know that I've mentioned their use several times, but it must be remembered that an adapter represents another mechanical connection in the feedline that contributes to the total system loss. You might wish to consider, when possible, avoiding the use of an adapter by replacing the incorrect connector on the cable with one that correctly mates with your scanner.

Preamplifiers & Converters

For reasons stated above in regard to losses caused by mechanical connections, converters and preamplifiers should be used with some knowledge of what's involved.

Converters are often determined to be necessary when you want to receive frequencies for which your scanner is not otherwise equipped. Right now, one popular converter, for instance, is available that permits reception of 806 to 912 MHz on any scanner that wasn't designed to pick up those frequencies, but which can receive the 406 to 512 MHz standard UHF/UHF-T bands. When not in use, the converter may be turned off and kept "in line" so that normal scanner reception on other frequencies can take place. While it's true that there will be only a slight signal loss resulting from it being left in-line when not in use, nevertheless it is a loss, no matter how negligible. If you aren't going to be using the converter for any extended period, it might be just as well to remove it from the antenna system (and take out the battery, too).

Preamplifiers are misused and abused within the monitoring hobby. Usually, you should never or rarely have a need for one, and when you do it belongs mounted right at the base of your antenna and NOT at the antenna input of your scanner. The purpose of a preamp is to amplify **signal** and boost it over **noise**. A preamp installed down in the radio room amplifies only what goes into it, and that includes weakened (by the coax) signal, plus noise picked up and created by the coax itself. A preamp installed at the base of the antenna won't amplify noise in the coax, but will amplify signals before the enter the coax. Down at the other end, signals and noise will be weakened by the coax, but the signal will still be over the noise by a greater margin. This is the sum and total key to successful monitoring and reception-- the signal-to-noise ratio (S/N). It's always good to remember this.

There are other things you'll want to know about preamps. At VHF frequencies (above 100 MHz), man-made noise grinds to a standstill and just doesn't exist. Noise above 100 MHz comes from several sources including the cosmos, the sun, and matter. Yes, matter... you know, "stuff." All matter produces RF from atomic and molecular activity. Some stuff generates more noise than other stuff. Below 50 MHz, the design of preamplifiers isn't too critical, and just about anything will work to amplify signal over noise. Not so at VHF/UHF frequencies.

Transistors which amplify very well and generate very very little noise at frequencies below 50 MHz start to freak out at 100 MHz and up. VHF and UHF preamplifiers are very critical with respect to design and choice of components and materials. Not just anything will do.

There's another characteristic of preamplification that you'll want to know about: **As bandwidth increases, gain decreases and noise increases.** This means that it is difficult to make a low noise, wideband preamp with good gain for the 25 MHz to 1300 MHz portion of the spectrum. I'm not saying that it's impossible, just awfully tough. So far as I know, no off-the-shelf product currently available in the scanner market can do the job with all of the stops pulled out. The specs of a proper scanner preamp I'd like to have at my station and yours are shown in Table 2-5.

Table 2-5

SPECIFICATIONS FOR A WIDEBAND PREAMPLIFIER

TERMINOLOGY	IDEAL SPECIFICATION	ACCEPTABLE SPEC
Noise Figure:	Less than 1.0 dB	Less than 2 dB
Gain:	Greater than 15 dB	Greater than 10 dB
Bandwidth:	25 Mhz - 1300 MHz,	50 Mhz - 1000 MHz
1 dB Compression:	Greater than 20 dBm	Greater than 12 dBm
Other:	Weatherproof, ruggedized for outdoors	No change
	Price: Less than $150.00	$200.00

The Noise Figure of a preamplifier is so very important at VHF/UHF frequencies that it can't be overemphasized. There would be little benefit in putting a preamp in your system that contributes as much noise of its own as it amplifies signal, and I've seen several scanner preamps that do just that! Those that shine in the noise and gain departments universally have another problem with the "1 dB compression" specification. This one is too technical to get into here, but what it means in layman's lingo is how much intermod will be added to your scanner when you turn the preamp on. The bigger the number, the better. Superior commercial preamps can be rated at 20 dBm, so 12 dBm as suggested in Table 2-5 should be a goal for the designers.

As a rule of thumb, don't spend good money on a scanner preamplifier. It will probably do you little or no good in a big city environment where hundreds of strong signals can invade the preamp and emerge as nasty, disagreeable intermod to effectively shut down your monitoring. A rural environment where there are few signals of any kind, much less strong ones, can be an excellent place to use a preamp. But even then, the specifications given in Table 2-5 are important. Get one as close to those specs as you can find.

The priority of specifications depends on whether you're in a metro area or in the boondocks. Our priority suggestions are shown in Table 2-6.

Table 2-6

IMPORTANT SPECIFICATIONS FOR A PREAMP

PRIORITY	RURAL ENVIRONMENTS	BIG CITY ENVIRONMENTS
1	Noise Figure	Noise Figure
2	Bandwidth	1 dB Compression
3	Gain	Bandwidth
4	1 dB Compression	Gain
5	Price	Price

Note that in both environments, **gain** isn't the prime consideration. If a preamp offers anywhere between 7 and 19 dB gain, that will be sufficient-- you don't need lots of gain. The variables of importance are Noise Figure, Bandwidth, 1 dB Compression, and Price. I put "price" at the bottom of the list because there are no free lunches when it comes to electronic equipment and you usually get what you pay for. You can factor it in at a higher position if you wish. For the "ideal" hobby preamp we described earlier, it's really a matter of "beggars can't be choosers," anyway because that unit doesn't seem to exist presently at any price.

Besides, the need for a preamp can be cut out or reduced by careful attention to loss-budgeting in the antenna and feedline, as I have outlined.

Tips & Hints on Scanner Receivers

Later in this book we'll get into the many neat things that can be done <u>to</u> your scanner. This section discusses some of the things <u>about</u> your scanner that are worth knowing.

For instance, have you you even given any serious thought to the features you need (or even want) in a scanner? VHF/UHF radios are just like cars in that they come in all sizes, makes, models, and performance ranges. You can't get around it. There's a range of choices of all these variables in the scanners made today. Normally, size and price are variables that we rather gracefully accept after the features and performance criteria have been selected. They're really the last thing to be factored into a decision. Size, however, can be a very different factor if we group all VHF/UHF radios into two classes: Bases and Handhelds. Certain handheld scanners can have all of the features and performance of certain base station scanners, but the handhelds may be smaller and less expensive.

There is really one additional classification of scanners that merits attention: <u>crystal controlled</u>. Scanners that require the use of plug-in crystals (one per channel) are no longer as popular with the public as keyboard programmable scanners. There are still a few models on the market, but most hobbyists would find little justification to buy one. With the crystals selling for about $5 a pop, you'd have to add $50 to the price of the scanner in order to monitor ten channels. A ten-channel programmable will cost less than an equivalent crystal scanner. The market for 2-to-4 channel pocket or handheld crystal scanners appears to be primarily volunteer firemen, off-duty law enforcement officers, and for certain business or industrial purposes. But, on the whole, these units no longer are seen in use for general monitoring and should eliminated from any serious consideration for such purposes.

However, when measuring yourself for a new scanner, you do have much in the way of features and performance to think about to get the unit best suited to tour requirements. You can add in size and price where you think most appropriate.

Table 2-7

LEGEND FOR TABLE 2-8

```
#  =  quantity                                      y  =  Yes or No response
>  =  more, greater or larger is better             +  =  Yes is better
<  =  less, fewer or smaller is better              -  =  No is better
?  =  subjective value; buyer must evaluate
```

Table 2-8

FEATURES, PERFORMANCE SPECIFICATIONS & VALUES FOR EVALUATING VHF-UHF SCANNER RADIOS

FEATURES	VALUE
Number of Frequency Bands	# >
Frequency Coverage of Each Band	# >
Total Frequency Coverage	# >
Number of Memory Channels	# >
Number of Memory Channel Banks	# >
Limit Search (Programmable Low & High Limit Search)	y +
Direct Search (starting at any indicated programmed frequency)	y +
Number of Limit Search Banks	# >
Number of Search Memories/Channels (temporary storage)	# >
Speed of Scan and Search Functions	# >
Selectable Speed of Scan and Search Functions	y +
Priority Channel	y +
Assignable Priority Channel	y +
Number of Priority Channels	# >
Delay or Pause (Selectable) Before Scan & Search Resume	y +
Lockout Selected Channels	y +
Lockout Review (manual review of locked out channels)	y +
Display (quantity & type of information in the readout)	? >
Display (Lighting & visibility; quality)	? >
Manual Override of Scan	y +
Mode Selection (manual control of AM, NFM, WFM or SSB)	y + >
Search Steps (User selectable)	y +
Search Step Selection Choices (5, 12.5, 30, 50 KHz, etc.)	y + >
Search & Store (Automatic storage of frequencies found in Search)	y + >
Fine Tuning/Center Tuning for Search Mode (Automatic or Manual)	y + >
Center Tuning Meter (or Other Capability to Assure Center Freqency)	y + >
Memory Retention (when power is disconnected)	y +
Low Battery Reminder, Memory (Memory Backup) (Acoustic or Visual)	y +
Low Battery Reminder, Power Source (Handheld & Portable Scanners)	y +
AC and DC Powered (Base and Mobile)	y +
Tape Recorder Output	y +
Signal Meter (S-meter)	y +
Headphone/Earphone Jack	y +
Mobile Mounting Bracket	y +
Keyboard Lock Switch	y +
Computer Interface (RS-232)	y +
Weather Alert	y +
Potential for Operator Modifications (Capability of User Upgrade)	? >

PERFORMANCE & SPECIFICATIONS	VALUE & UNIT
Sensitivity on each Band and AM, NFM, WFM, SSB	< uV @ 20dB (S+N/N)
Signal to Noise Ratio; each Band and AM, NFM, WFM, SSB	> dB @ 100 uV
Selectivity: AM, NFM, WFM, SSB	< ± KHz
Image Ratio	> dB (decibels)
IF Rejection	> dB (decibels)
IF Conversion (# of IF frequencies)	> 1, 2, or 3
Modulation Acceptance	> ± KHz
Spurious Rejection	> dB (decibels)
Squelch Sensitivity	< uV (microvolts)
Residual Noise	< mV (millivolts)
Priority Channel Sampling Rate	? seconds
Scan Delay Time	? seconds
Audio Output Power	> watts
AC Power Consumption	< watts or va
DC Power Consumption	< watts or va
Weight	< lbs or kg
Price	? <

A gel-cell battery pack and trickle charger. The pack consists of two 6-volt gel cells wired in series for 12 volts. The charger can run 3 to 10 amps.

Table 2-8 lists many of the features and performance specs you will have to select from in choosing the radio of your needs and desires. It will be helpful to understand that "features" are the accoutrements or functions that come with the scanner. Performance specifications are a measure of how much relative value those features have in a scanner, and how well they work. Table 2-7 provides a legend to assist you in perusing Table 2-8, which lists those features and performance specifications along with a "value" list that tells you what to look for and whether "more" or "less" is ideal.

There are many ways to use Table 2-8 to help determine the best scanner for you, or the best on the market. One way would be to assign a "point value" (1 to 10) for the relative importance of each item. Then, compare all scanners in which you are interested against Table 2-8. Add up the totals and your choice might be made a lot clearer. This information, at the very least, can prepare you to be an informed buyer.

Emergency Power For Your Scanner

Introduction: A scanner might well enhance your chances for survival or well-being in a disaster, crisis, or emergency situation. It is precisely during such times that the 117 volts of AC that normally pours forth from the sockets in the walls of your home may suddenly become either unreliable or totally nonexistent. DC power from battery packs, including automotive batteries, can operate most scanners even if they weren't designed for battery operation. If you are prepared with at least a DC power cable and (better still) a well designed, well thought-out battery pack, then you might be able to retain contact with the outside world in a time of emergency. Being prepared might spell a difference for you, your family, and/or your community. Those who survived Hurricane Hugo in South Carolina, or the October '89 San Francisco/Oakland area earthquake know just what I mean.

DC Power Requirement Analysis for the Realistic PRO-2004. The manufacturer's specs for the PRO-2004 scanner state that the AC power requirements are 120 VAC @ 60 Hz and that the power consumption will be 20 watts (maximum). The DC requirements are stated to be 13.8 VDC with a power consumption of 12 watts (maximum). This would imply that the the current drain could be as high as 0.870 amperes (870 ma). Fortunately, minimum power requirements are not nearly that high.

I've long been interested in "survival communications" and emergency services, so I wondered just how my Realistic PRO-2004 would behave under different and uncertain conditions of DC power. So, I set up my DC power supplies and test equipment to determine what could be counted on in terms of performance when power sources (batteries, etc.) were at a premium. Table 2-9 shows (surprisingly) what was discovered:

Table 2-9
DC POWER REQUIREMENTS FOR PRO-2004

DC POWER SUPPLY VOLTAGE (DC)	CURRENT DRAIN AM/NFM	WFM	POWER CONSUMPTION (WATTS) AM/NFM	WFM
13.80	400 ma	443 ma	5.5	6.1
12.00	385 ma	433 ma	4.6	5.2
11.00	375 ma	413 ma	4.1	4.5
10.00	338 ma	370 ma	3.4	3.7
9.00	300 ma	328 ma	2.7	3.0

Wide FM mode draws about 10% more current than other modes.

Wow! The tests disclosed several things of great interest to those who might want to power their PRO-2004 (and probably the PRO-2005) from a DC source. I should point out that my PRO-2004 is loaded with modifications, some of which draw a bit of power. Chances are that a stock PRO-2004/2005 will draw less current than shown in Table 2-9. Here is a summary of my findings.

A. Maximum DC power consumption is 6.2 watts; minimum is less than 3 watts.

B. The PRO-2004 operated normally in all respects as the DC supply voltage was slowly decreased to 8.60 volts. Audio volume (loudness) was noted to drop slightly as the supply voltage was decreased, but that effect is hardly worth noting.

C. At 8.60 volts, the digits in the LCD display blanked out and the radio stopped operating, even though the backlighting continued to function.

D. As the voltage from the DC power supply was slowly increased starting at 8.60 volts, nothing happened until the voltage reached 9.20 volts at which time, all scanner functions came back on and were normal.

E. With the PRO-2004 turned **off** and with a range of supply voltages between 8 and 9.20, the unit would not operate when turned **on**, unless the supply voltage was 9.20 or greater

Conclusions: The Realistic PRO-2004 scanner can serve as a near-ideal emergency communications scanner because of its exceptional capability and low, non-critical DC power requirements. It will operate normally in all respects even with the DC supply voltage dropped below 9 volts. Once turned off, the unit requires a minimum of 9.20 VDC to start up. After startup, the scanner will function normally again so long as the voltage doesn't dip below 8.60 VDC.

This is an ideal situation for portable, emergency power situations because the supply voltage can be designed for a range from, let's say 9.5 VDC to 15 VDC. A battery pack made of nine or ten flashlight "D" cells wired in series will readily power the PRO-2004 for a time! Ten or eleven nickel-cadmium cells wired in series will make a rechargeable power supply. Other types of handy rechargeable power packs could include one 12 VDC "gel cell" or two 6 VDC "gel cells" wired in series. Even a solar power supply could be pressed into service since nominal power requirements can be stated to be 12 VDC @ 0.5 amps (500 ma). Solar cells are expensive, though a one-time investment since they don't wear out or run down. A solar panel designed for 12-14 VDC @ 500 ma could be used to charge nickel-cadmium or gel cells during the daytime in order to power the scanner by night. Two such solar panels will keep the batteries well charged and still operate the scanner in the daytime, too. (For maximum effectiveness of recharging batteries, the solar panel should be designed for 13.5 VDC @ 500 ma. The foregoing discussion and tests should provide ample food for

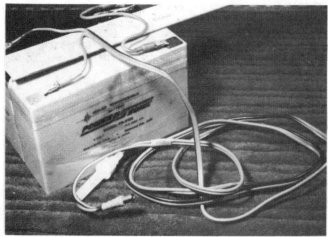
Typical gel-cell power pack and cable for scanners. The power cable is fused for 1-amp. The plug in the photo is for the PRO-2004/2005.

thought for survivalists and others who realize that their PRO-2004/2005 is worth its weight in gold during a crisis or emergency.

One reason the PRO-2004 (and probably the 2005, and most other modern scanners) is so flexible with respect to DC power requirements is because (with one exception) nothing in the radio operates directly from the specified 13.8 VDC-- we'll get to the one exception in a moment. The PRO-2004 is chock full of "voltage regulators" designed to drop that 13.8 VDC down to 5 to 8 VDC. A voltage regulator is specially designed to accept a wide variation of input voltages and still maintain a constant, steady output voltage. This is why the PRO-2004 will operate reliably down to as low as 8.6 to 9.2 volts! You really don't need 13.8 VDC, because anything between 9.5 and 15 volts will do nicely. I really don't know what the upper DC voltage limit is, and I don't intend to find out either. But we do know that the PRO-2004 was designed to be powered by automotive electrical electrical systems which, when charging, can be as high as 15 volts. My guess is that the PRO-2004 will accept voltages as high as 16-20 volts DC before something blows, but you can't go wrong if you set 15 volts as the absolute maximum, tops, not to be exceeded.

Only one circuit in the PRO-2004 is powered directly from the +13.8 VDC external source (or from the 110 VAC internal +14.0 VDC power supply). This is the Audio Power Amplifier (IC-7), which drives the speaker or headphone, but is not critical with respect to supply voltage. IC-7 can operate over a range of at least 8 to 16 VDC. Audio power will decrease a little as distortion increases at lower supply voltages, but neither enough to affect normal listening. Other circuits in the PRO-2004 are powered from one of several voltage regulators which provide a steady 5 or 8 volts regardless of the input voltage-- which we now know can range from 9.2 to 15 VDC.

Make a DC Power Cable for The PRO-2004/2005

If you ever want to power your PRO-2004/2005 from a DC source, just be sure that correct polarity is observed in the coaxial DC power plug! The center pin is (+) hot while the shell or outer ring is (-) ground. Radio Shack carries a preassembled power cable (270-1534) with a fused car cigarette lighter plug on one end and the proper DC power plug on the other that mates with the "DC 13.8v Jack" on the rear of the PRO-2004/2005.

You can also fabricate your own DC power cable using a pair of insulated (16-20 ga) wires of your choice and a 5.5mm x 2.5mm coaxial DC power plug (Radio Shack 274-1573, or equivalent). If you fabricate your own cable, it is strongly suggested that you install an "in-line" fuse holder (Radio Shack 270-1281) and a fast-acting 1-amp

Hombrewed 13.5 VDC alkaline battery pack. Two four-cell packs are glued together and rewired to make a nine-cell (13.5 volt) pack.

fuse (270-1273) in series with the hot (+) lead of the power cable. If the 1-amp fuse pops for no apparent reason during testing, then go to a 1.5 amp fuse (270-1274). You should install the in-line fuse holder as close to the coaxial DC power plug as possible.

Emergency Power for Other Scanners

Even if you do not own a Realistic PRO-2004/2005 scanner, please refer to the foregoing comments, anyway. The contain general information and concepts that need not be repeated here.

Other modern scanners will probably compare favorably with the power requirements determined for the PRO-2004. Digital and analog circuits in most electronic equipment require steady, regulated +5 and/or +8 VDC, so your scanner will undoubtedly accept a range of input voltages. Your scanner will be designed with appropriate internal voltage regulators. When in doubt, consult with the specifications and schematic diagrams in the Service Manual for your scanner. You can also perform tests to determine the unit's **minimum** DC power requirements. Here's how to do this:

Determine the **rated** DC supply voltage (see Service or Owners Manual) for your scanner, then connect a variable DC power supply preset to that rated voltage to your scanner. Typically, this will be +13.8 VDC, but it could be different, especially for handheld scanners. Important: Observe correct polarity! Connect a DC voltmeter to the power supply terminals. Connect a DC milliammeter or ammeter in series with the (+) lead from the power supply to the scanner. Turn the scanner **on**; find a station; adjust the volume control to a comfortable, normal listening level; then observe the current drain as indicated by the ammeter. Operate the scanner in its various modes and settings to determine the highest and lowest current drains. Write them down along with the measured supply voltage.

Now, gradually reduce the voltage from the power supply by a half-volt or so at a time, observing and testing the scanner for proper operation. When the scanner begins to operate in an erratic manner, or it ceases to operate altogether due to the lowered supply voltage, slowly increase the supply voltage until proper operation is resumed. Write down <u>that</u> voltage and the corresponding current drain.

These figures will be the **minimum** DC power requirements for your scanner. Do **not** perform a similar inquiry to find the maximum. When you know the minimum, however, you can you can then design an "emergency power pack" based upon those figures. **Do not ever operate any scanner at a DC voltage more than 10% above that specified by the manufacturer!** This will normally be 13.8 VDC for scanners that are designed to be operated from an automobile, thus the 10% overrange will be 15.2 VDC. Note that handheld scanners typically require 6 to 9 VDC, and again, the 10%

These assorted power cables show some of the several ways you can feed power into your scanning receiver. For maximum versatility, prepare a cable for every potential method you might ever need to use to power your scanner.

overvoltage warning applies. Therefore, most handheld scanners cannot be directly powered from car cigarette lighters unless a special voltage reducing power adapter is used.

No harm can come from **under**voltage other than possible loss of scanner memory. Even that is unlikely, especially if there is an internal battery for memory backup. Undervoltage for brief periods of testing should not disable the memory in any event.

Battery Eliminator Power Supplies for Your Handheld Scanner

More than likely, your handheld scanner uses six "AA" alkaline or nickel-cadmium cells. Batteries are an excellent source of mobile and temporary power. In fact, it's about the only way to go for these situations. But batteries run down, and even nickel-cadmium cells "wear out." Pity this could happen at a time when you might need your scanner the most!

Batteries are somewhat like oil, gas, coal, and minerals; non-renewable resources! You think they're renewable? You say that all you have to do is run down to the store and buy another set any time you need them? Sure you can. But a disaster-- local, regional, or national-- could change that "renewability" in the twinkling of an eye. In the aftermath of an earthquake, hurricane, flood, power blackout, or other similar problem, try and buy a set of alkakine "AA" batteries. You won't be able to do so with a crisp $100 bill, even if you whistle three choruses of "Dixie" as an added enticement.

Even during the routine and regular use of a scanner, the cost of replacing, and or the hassle of constantly recharging nickel-cadmium cells can become tiresome. If you use a handheld at a fixed location with a handy source of AC power, or in your vehicle where 12-14 VDC is readily available, there are convenient alternatives to unhook you from being completely dependent upon "AA" batteries.

Fortunately, you can save the alkaline and nickel-cadmium cells for when you need them most, but most of the rest of the time you can use other convenient methods to power up your handheld. If you're a serious scanner user, you should have these alternatives at your disposal, anyway. Radio Shack makes 9 volt AC-DC converters for use where AC power is available (273-1455 or 273-1650). Just plug the power unit into a wall outlet, and the other end into uour scanner's external power jack. For vehicular use, you'll need a DC-DC adapter that converts 12-14 VDC to 9 VDC. Radio Shack has two units that do this, 14-844 and 270-1560.

It makes sense to acquire and learn to use other sources of power for your scanner. Having a variety of low cost proper adapter units and cables for a variety of power sources can save money in the long run and enable timely use of your scanner during any emergency or crisis.

Emergency 12-14 VDC Power for Your Handheld Scanner. If you are in an emergency situation and the batteries in your scanner are dead, and if you don't have a DC-DC power adapter for converting 12-14 VDC to the voltage required by your scanner, you can still (as a last resort) get emergency power from an automotive electrical system. To be candid, there could be some risk to this scheme, depending upon the power requirements of your scanner, but in a real pinch you'd have little choice. Here's what you do:

For Negative Ground Electrical Systems Only: Most modern American-made vehicles have negative ground electrical systems-- you may wish to confirm this for your own vehicle by looking at the vehicle's manual. Prepare two wires for connection to the power terminals of your scanner. Designate one wire as positive (+) and the other as negative (-), and then connect them to the appropriate terminals. Connect the negative (-) wire to a paint-free location on the frame or chassis of the vehicle or to the (-) terminal of the battery. At the free end of the (+) wire, connect one end of a resistor of approximately 100 ohms, with a 1 or 2 watt rating. Actually, the resistor can be anywhere between 80 and 120 ohms.

Connect the free end of the resistor to the hot (+) side of the electrical system (12-14 VDC). The resistor will drop some of the excess voltage leaving 7 to 10 volts for the handheld scanner, which should work just fine within that range. The caution here is that the lower the ohms rating of the resistor, the higher the voltage to the scanner, and the scanner could blow up if the voltage is too high. There won't be any harm in using one with too high a rating in ohms; the scanner just won't function properly. Maybe, if the scanner has one, a "low battery" indicator might beep or light up to tell you that the unit is starving for voltage. Still, this method must be considered as being for emergencies only because there is an element of risk of damage to your scanner if you miscalculate or otherwise goof.

Tips on Nickel-Cadmium (Ni-Cd) Batteries

Rechargeable nickel-cadmium (Ni-Cd) batteries have become a way of life, and for equipment that is designed for Ni-Cd cells, there is little to know or be concerned with. The main thing for all Ni-Cd batteries is that they can develop a "memory" after a while if they are not fully discharged and then promptly recharged. This is to say that Ni-Cd batteries should be periodically discharged to the point that the equipment doesn't work, & then fully recharged before using the equipment again. If this ritual is not done regularly (like once every two weeks or maybe a month) then the Ni-Cd cells will not give the length or quality of service according to their specifications. Don't allow Ni-Cd cells to remain floating on a charger for lengthy periods without putting them through a discharge cycle.

These days, Ni-Cd batteries are used for situations and equipment where Ni-Cds weren't originally intended. In general, it's OK because voltage is voltage and current is current. If a scanner is supplied with voltage within its specs, it will work--- period. But Ni-Cd cells might not work as long as you would want or expect. If you use Ni-Cd cells for emergency power to replace standard alkaline batteries, then there are a few things you'll want to know about them.

A freshly recharged Ni-Cd cell will measure about 1.4 volts at its terminals, but will drop to 1.2 volts shortly after removal from the charger, or at least immediately as soon as placed back in service. This has significance if you design a Ni-Cd pack for a situation where it's ideal to maintain a voltage level for some period of time. For instance, if you need 12 volts for a period of time, ten Ni-Cd cells in series won't do the job. Even though the initial terminal voltage will be 14 volts, it will drop rapidly to 12 volts, and then under use, the voltage will steadily drop from 12 volts down through the discharge cycle.

In summary, Ni-Cd batteries don't come with high recommendations for anything except cordless telephones, flashlights, portable hand vacuum cleaners, electric

Chassis grounding strap for a base scanner. A female spade lug facilitates a quick disconnect from the main ground strap. An alligator clip can also be used for this.

toothbrushes and razors, and other low demand, short-use equipment. I suppose that Ni-Cds are passable for handheld scanners, but you owe it to yourself to measure the complete operating time you are going to get out of a pack of freshly charged Ni-Cds. Mark the starting time and set the scanner on a single channel which is active some of the time-- then let it run until it quits. Again, mark the time and calculate how long it lasted. That's all the time you can count on, and probably less than that if and when a time comes that you really need your scanner!

Tips on Lead-Acid Storage Batteries (& Gel Cells)

Now here's the way to go, except that it's not a lot of fun to lug around a car battery for more than maybe a few seconds unless you're a physical fitness fan. Fortunately for the rest of us, car batteries aren't the only type. There are several kinds of lead-acid batteries other than those made for cars, and all possess some of the same general characteristics at 2.1 volts per fully charged cell.

The typical "12 volt" lead-acid battery consists of six cells, so the terminal voltage at full charge is 12.6 volts. Just like Ni-Cd cells, lead-acid cells actually have a temporarily higher terminal voltage immediately after a full charge. This voltage will be 13.8 volts if float or trickle charged, and 14.4 volts if charged to capacity. Automotive charging systems typically generate 14.5 to 16 volts. The instant the engine is shut off, the battery voltage will be about 14 volts. Within a short time, it will "settle" back to 12.6 volts.

Lead-acid batteries generate and maintain a lot of power for their volume; more so than any other kind of battery. This makes them emminently suited to communications applications. Despite their weight per volume, some lead-acid batteries are small enough to be convenient and still pack useful power. The computer revolution helped create a class of sealed, maintenance-free battery called a "gel cell" of which there are ample quantities on the commercial surplus market now.

Varying in size from about a "D" flashlight battery to cigar box sized or even larger, gel cells possess all of the good characteristics of the common automotive battery and fewer of the bad qualities. Gel cells are rechargeable with conventional equipment (automotive trickle chargers) and they generate adequate power for reasonable periods of time. Also, they come in all voltage and current ratings. Those most ideal for use as scanner emergency power sources would be rated at 6 or 12 volts and at about 10 ampere/hours (AH). Two 6 volt gel cells can be connected in series to provide 12 volts. Two or more 12 volt gel cells cal be connected in parallel for additional ampere/hour capacity.

All 12-volt gel cells should be recharged at 14.4 volts until fully charged, and may then be left indefinitely on a trickle charge of 13.75 volts. If left on 14.4 volts too long, the cells can be damaged. The max charging current should not be more than 25% of the ampere/hour rating of a single battery in the pack. Two 6 volt gel cells can be left connected in series and charged with normal 12 volt charging equipment. If charged separately. 6 volt cells should initially be fully charged to 7.2 volts and then trickled with 6.88 volts. Again, the maximum recharge current should not be greater than 25% of the ampere/hour rating.

Tips on Alkaline Batteries

Alkaline "AA" cells are commonly used to power handheld scanners. Service life is good and many hours of operation can be expected from a fresh set of batteries. Most handheld scanners use six alkaline batteries to provide a starting voltage of 9 volts which remains fairly constant for some time. The voltage of the pack does slowly drop during use to as low as 4.5 to 6 volts before the scanner quits.

Some people tell you that alkaline batteries can be recharged. I have tried it with mixed results and conclude that the procedure is not worth doing. An elaborate setup is required and there is always the danger of the batteries exploding or damage occurring to the equipment. Moreover, the recharge runs out very quickly. Forget this idea; just keep a spare set of alkaline batteries on hand for emergencies.

A very handy source of emergency power for your handheld scanner can be put together with the same number of "D" or "C" cell alkaline batteries as the number of "AA" cells normally used, which is typically six. All you need are the batteries, battery holders, a length of paired wire and a suitable power plug to mate with the "external power jack" of your scanner. "D" or "C" alkaline batteries have a <u>far</u> greater service life than "AA" batteries, and would be ideal for crisis or emergency situations where you want to have a power source providing both time and reliability. The battery holders, wire, plus the appropriate power plug are available from Radio Shack. There are two precautions to observe:

1. You must first determine the polarity of the "external power jack" on your scanner. Radio Shack handheld scanners have the tip or center of the jack negative (-) and the shell or ring as positive (+). Other scanners may be different, including Radio Shack's base scanners where the tip of the "DC power jack" is positive (+). Check your owner's manual. Some scanners even have the polarity information shown on the case right next to the power input connection.

2. The battery holders must be wired in series. If your handheld scanner uses six batteries, you can use six single battery holders, or one that holds four plus another that holds two, or three holders that accomodate two each. Just be <u>sure</u> to wire them in series, observing the correct polarity (+ to -, and - to +) . When <u>done</u>, there should be two wires left free, one negative (-), and the other positive (+), which should be connected in proper polarity to the correct power plug.

Operating Hints & Techniques

There isn't very much even an expert and experienced scannist can tell you about the operation of your scanner. Information on what the knobs and buttons do is all in the manual that was supplied with the unit, so there wouldn't be much point here in explaining how to set a squelch control or why to lock out a channel. Instead, I thought I'd pass along some thoughts on tools and general approach. As a former educator, I learned that it is enormously more valuable to the recipient to convey methods, procedures, strategies and tactics than hard facts and things to commit to memory. I'll attempt to do as much here; to assist and enourage you to take up where this leaves off; to open a door for you.

Increase Your Knowledge: VHF/UHF monitoring is one of those pursuits where the more you know, the more enjoyabe and worthwhile it becomes. The essential knowledge comes from several potential sources, including:

A. Experience-- there's nothing like it and nothing exactly takes its place. It comes with time and intensity of self-application. It may take a while to acquire, but it's worth having. On the other hand, it's easy and fun to get and it's gained as you participate and explore monitoring.

B. Books and Periodicals - also hard to beat. Knowledge gained from reading and study, and then put into applied use becomes an express train to experience. Periodicals and reference books go a long way towards connecting the scannist with the "big picture" of the monitoring hobby. The mistakes, successes, experiences, and valuable opinions of others, as well as new trends and products, are conveyed via this media.

C. Association with others - direct contact with persons of like mind and interests can help the scanner hobbyist gain a strong foothold on knowledge and experience. There's nothing like enjoying a hobby in conjunction with one or more kindred souls.

The first and most important tip or hint that I can convey to you is this: Dedicate yourself to the patient acquisition of knowledge about your hobby. All other tips, hints, and techniques in this book are secondary!

Obtain Frequency Directories. Granted, you can probably set your scanner to the **search** mode and in a day's time you'll have discovered hundreds of active frequencies. But that's all you'll have, and after a few days all of those raw numbers will make your head swim. For the most part, there will be little organization, rhyme or reason to all of those numbers you've collected. Random frequency **search**ing is akin to a police officer searching for a stolen car by standing on a downtown streetcorner and writing down the license plate numbers of every vehicle that passes by. So you have a list of numbers-- now what?

If you've monitored the VHF/UHF frequencies, you may have noticed that few, indeed, are the stations that bother to properly identify themselves by their assigned call letters or name of their operating agency or company. If you patiently monitor a frequency long enough, there will be certain things you can deduce from what is communicated as to what function the station performs, such as police, fire, medical, business, road repair, power utility, highway assistance, taxicab, etc. You might even be lucky to catch a callsign or other usable identification for your efforts, but this approach is going to involve a lot of work to find out something that you could look up rather quickly in a published reference source intended as an aid to scanner users.

Frequency guides come in all sizes, shapes, and coverages. Some zero in on specialized topics or interests, such as Tom Kneitel's popular **Top Secret Registry of U.S. Government Radio Frequencies,** or his **Air-Scan Directory of Aeronautical Communications,** as well as his **National Directory of Survival Radio Frequencies.** Then, there's the American Radio Relay League's (ARRL) **Repeater Directory.** Other directories are wider in scope but are for for specific geographic regions such as the seven different volumes of the **Uniden/Bearcat Regional Frequency Directory** series. Others still, are directed to specific states or local areas, such as the **Scanner Master New York Metro/Northern New Jersey Guide,** by Warren Silverman; the **Official Massachusetts Scanner Guide,** by Bob Coburn; the **San Diego Scanfan,** by K.C. White, to name only a few out of dozens.

There is no single "ideal" or "most comprehensive" directory. Each serves a specific and sometimes unique purpose for a target group of monitors. As your interest in VHF/UHF monitoring grows along with your experience, so should your reference library of frequency directories. For starters, a basic library of several frequency guides will go a long way towards helping a person systematize their growing hobby interest. Then, as more information and experience is gained, additional directories can be added for guidance in specialized areas of interest. Frequency directories often contain a lot more information than station callsigns and locations, many also provide system data, signals and codes, maps, as well as incidental information about scanning, including tips, hints, and kinks.

Subscribe to The Scanning Hobby Periodicals. One of the best ways to keep abreast of

You'll find that scanner-oriented publications enhance your knowledge and enjoyment of communications.

the latest equipment, supplies, frequency guides, techniques, and discoveries about the monitoring hobby is to read the monthly magazines devoted to monitoring. There are two magazines of some size that have been around for several years covering monitoring in depth, these are:

Popular Communications Magazine
76 North Broadway
Hicksville, NY 11801

Monitoring Times
P.O. Box 98
Brasstown, NC 28902

Both are excellent and each is sufficiently diversified to contain interest everyone. Popular Communications is sold by subscription and is also available nationally on newsstands.

You'll find that the hobby publications will keep you current on art of scanning.

Be in Contact With Others. There are so many people with scanners these days that you undoubtedly work or go to school, or church, or bowling, or hunting with other local people who are into scanning and would like nothing more than to swap notes and share frequencies and experiences with you. You can meet a lot of these folks at communications shops (as customers or sales personnel), or maybe standing near the runway at any airport, with a handheld scanner watching the planes take off and land. Don't be bashful, most of these people welcome inquiries about monitoring and are quite happy about hooking up with other local people with whom they can share their most exciting hobby. The mutual exchange of information benefits all concerned.

There are also classified ads in the hobby magazines from scanner fans seeking local contact with others of their ilk, or wishing to exchange tapes or letters with distant monitoring fans who may share their interest in a particular specialty area of the hobby.

There are also scanner clubs, some regional, some with members throughout North America. These groups usually issue a monthly newsletter or other publication directed at their memberships. Usually these publications contain frequency listings as well as tips and techniques. Some of these scanner club publications are quite good, just as others don't really have much to say of value. Clubs tend to come and go, and also their newsletters give off the (possibly false) impression that they can get a bit snooty, vindictive, cliquish, petty, or persnickety from time to time, and with very little apparent justification. Because these clubs are a lot like people, unique and individualistic and with a distinct personality of their own, let's merely suggest that you might wish to check with other scannists to see which (if any) hobby groups they belong, and if they feel that membership was beneficial. When you collect some good opinions, also collect the groups' names and addresses and seek out membership info.

Build Your Hobby Library. Frequency directories, monthly periodicals, and perhaps the newsletters from any groups to which you belong, should comprise major elements of your library, but you'll want to include information on antennas, theory, and other related aspects from various sources. The number of good titles and subjects are too numerous to list here, but you may want to contact the following publishers of books relating to scanning to ask for their catalogs:

CRB Research Books, Inc.
P.O. Box 56
Commack, NY 11725

Grove Enterprises
P.O. Box 98
Brasstown, NC 28902

Official Scanner Guides
P.O. Box 712
Londonderry, NH 03053

Tab Books, Inc.
Blue Ridge Summit, PA 17294-0840

Of course, many local and mail order dealers stock books from these as well as other publishers. Check the ads in Popular Communications and Monitoring Times for the names and addresses of communications equipment dealers offering catalogs.

Organize & Take Pride in Your Monitoring Station

If the extent of your monitoring interest is to stick your scanner on a shelf and then idly listen to a favorite police or fire channel when there's nothing to watch on TV, then this section is probably not going to offer you much benefit. On the other hand, if VHF/UHF monitoring offers you a fair measure of relaxation, enjoyment, and anticipation, then you owe it to your increased enjoyment to allocate some space and household resources to properly support your activity. Look at it this way, even your dog has a favorite spot all his own that he trots off to for those times when he wants to relax, be it a doghouse in the back yard or under the couch in the den. The kids have their playroom, the cat has his favorite window sill in the sun. Who, then, is to deny you at least a corner of some room just for radio monitoring purposes?

It really doesn't take much space to properly support the hobby of scanner monitoring, but it does take a little something in addition to the scanner itself. You don't necessarily have to have an entire, official "radio room" like some people, but you do need something definite in the way of a designated area. Perhaps a corner of the den or family room, a corner of the garage or basement, even a large closet. If you devote several hours a week or more to scanning, then the minimum things you need can be suggested as follows: 1. Desk or small table; 2. Chair; 3. Shelf; 4. Drawer.

Comfort, a little elbow room, and facilities for some organization are the keynotes to these suggestions. The table or desk offers room to lay out reference materials, log sheets, a clock, lamp, maybe a tape recorder, and the scanner itself. The space can also be available for you to pop the case off the scanner and do a little work inside.

The chair is for comfort and ease of work. The shelf is to hold your books and other reference materials not in use. If the shelf is at eye level, you might even put the scanner itself there. The drawer is for unsightly things such as tools, scraps, pens, rulers, etc. Get the idea here? Organization and comfort are the two vital ingredients to pleasure and success in anything.

Organization. Scanning can be as large or small a pursuit as you wish it to be. If it captivates you, if you enjoy it, and if it fills a regular portion of the time available to you, then it will undoubtedly gow on you with some rapidity. Your progress might not be measurable over days and weeks, but after a year or two goes by and you take stock of your progress, you'll see where you've come a distance. Now, get this: there's no limit to how far you can progress; the only limit is you. Well, there are some other limits, but they are under your control.

One major limitation is or will be your **system of organization** that supports your growth in the hobby. If you and your scanning materials are organized, your progress will be much faster than others who pay little heed to organizational concepts and drift around aimlessly in a myriad of different directions. A notepad and pen or pencil are the first tools you'll need to become organized. Better still, get a 6" x 9" 3-ring binder filled with looseleaf note paper.

Keep a sort-of diary; it need not be anything very complex. Use the notebook to jot down frequencies, ideas, questions, and things you discover during the pursuit of your radio interests. When you read the monthly magazines and find things of interest to you, jot down a notation or two along with the name and issue of the magazine so you can go back and refer to the information at a future date. If you become seriously interested in the scanning hobby, there is no way in the world that your memory will be able to assimilate, store, file, and recall at will all of the information, data, frequencies, and ideas that will come your way over a period of time. Take a few moments each day or week to highlight things in your notebook and before long, you will have built up a wealth of readily accessable information you can quickly draw upon, much of which would have otherise been forgotten or lost. There is something almost mystical, too, about writing things down. The mere act of writing reinforces the memory.

Of course, there is more to organization than writing. The concept of "a place for everything, and everything in its place" is worthwhile to remember at every monitoring station. You don't need to have much, but those things you do have should be organized, particularly your tools and sources of knowledge and information. By that, I mean your books, magazines, directories, notes, and scraps of information. This becomes _very_ important as a person progresses through the various stages from novice to expert. Without organization, you can make some progress but sooner or later you reach a modest plateau or peak that can't be exceeded. If you're happy and suited at that point, then relax, enjoy, and good luck to you. Chances are, however, that with a little organization behind you, a much greater peak can be achieved, and that only means more relaxation and enjoyment.

Be Proud of What You Have to Work With. Pride is a great motivator. I'm talking about the good kind of pride; pride of making the best out of what you have to work with; pride of accomplishment; and pride of self. That means make your monitoring post functional in performance as well as pleasurable in appearance. Perhaps your spouse isn't as enthusiastic about scanning as you are, but you'll hear a lot less about that lack of enthusiasm if the ol' monitoring post looks good, is neat, and doesn't detract from the room in which it is located. That's the least of it. The main thing is that _you_ will feel good if your station looks good and works well.

Work At It! Huh? Yeah, there's no pleasurable pursuit on earth that doesn't require a little work and self-dosciplene somewhere along the line. Scanning is no exception, though it's one of the least disciplined pursuits. But, believe me, there is a lot more to reaping the maximum enjoyment of scanning than just sitting there holding a hot cup of java or a cold Bud with one ear "tuned" to a fire or police channel while watching David Letterman on TV. That's the "couch potato" syndrome, and frankly, it's the least part of scanning.

There _is_ work to be done in really enjoying this hobby. There is a science, a mathematics, a physics, a mechanics, an art, and a knack to successful scanning. Face it, there are more than 200,000 frequencies in the scannable VHF/UHF spectrum (25 to 1300 MHz). At any given moment, most of these contain absolutely nothing of interest to anyone. The ones that do present something of interest are going to be "alive" one moment and silent the next. Without some knowledge, planning, science, and mechanics behind the process of scanning, the sounds of silence will prevail just as the chances of monitoring the rewarding communications are about on a par with winning the lottery.

Put a little science and mechanics into your efforts, with a dash of math and physics, and you get the knack-- you find that the sky becomes the limit. Instead of sorting through 200,000 frequencies without plan or purpose, you'll focus on less than 200, or maybe even as few as 20 for the really exciting comms. But _which_ ones will you focus on?

Scanner Frequency & Channel Management

Frequency management has received practically no attention within hobby books and magazines over the years. Most of the emphasis has always been on new equipment and the processes and mechanics of scanning. In fact, so little information has been presented on Scanner Frequency and Channel Management, that I had to resort to my engineering and business management background, and also enlist the aid of my trusty computer to develop the rudiments of a system to manage my scanner channels and frequencies.

How clearly do I recall my first digital, programmable scanner in late 1982, the Realistic PRO-2002. It had a whopping 50 channels and, for a year or so, I was in sheer ecstasy. Then it dawned on me that I was ineffectively utilizing the broader capability of all those channels. My venture to develop a system for managing that capability began in 1983 and has continued ever since. I don't profess to be an expert in this because I haven't "rubbed elbows" with others who may have worked in this area. My findings and procedures are largely self-taught, which does leave a lot of room for refinement and enhancement by others. I do propose to formally introduce this subject in print now and for perhaps for the first time. It's my hope that others will be interested and come forth to contribute to the pool of knowledge that is only lightly introduced here.

The basics of the science of scanning hark back to getting organized, and include frequency directories and "how to" books. For example, say you have a 400-channel scanner, like a PRO-2005 or a modified PRO-2004. There are many ways to program such sets, and most of them seen ineffective. If you haphazardly load up those 400 channels with random, unorganized numbers (frequencies), you'll hardly enjoy the results of your efforts. However, if you systematically employ the ten switchable banks of 40 channels each, using some rhyme and reason in an order of what was programmed, the results can be very meaningful.

For example, there is no point in running your index finger down the column of a frequency directory and programming each one that you identify as being from your geographic area. If you do it that way, the resulting bouillabaisse of boats, police, fire, medical, business, aero, government, and other frequencies will be a rather flavorless hodge-podge, and rather chaotic at that. If, however, you dedicate one 40-channel bank to law-enforcement; another to fire; another to local governments; another to federal government (a logical order and grouping), followed by a bank of aero, a bank of business, etc., then you'll have the ability to easily focus on specific interests as you desire. And, you still have the option to still run them all when you don't wish to zero-in on a particular grouping of stations.

Table 2-10 shows one possible way of how to dedicate and program ten 40-channel banks for a large metropolitan region:

```
                        Table 2-10
              PRACTICAL PROGRAM FOR 400 CHANNELS

BANK  CHANNELS   TITLE

 1:    1-40     Utility & Early Warning; state & local disaster response;
                  agency intersystems; REACT; weather & general miscellaneous
 2:    41-80    City Police & Law Enforcement Intersystems; Mutual Aid
 3:    81-120   Sheriff & Highway Patrol
 4:    121-160  Emergencies: Civil Defense; Red Cross, Search & Rescue;
                  Civil Air Patrol; Distress;
 5:    161-200  Medical & Paramedics
 6:    201-240  News Media; TV, Radio, Magazines & Newspapers
 7:    241-280  News Media continuation & Fire Departments
 8:    281-320  Military
 9     321-360  Federal Government
10:    361-400  Business, sports, rail & general miscellaneous
```

Table 2-10 isn't shown as being the "right" way to do this, it doesn't show special

Figure 2-1a
MASTER PROGRAM PLAN FOR 400-CH SCANNERS

PAGE_____OF_____
PROGRAM OR BLOCK No:____
DATE:_____

BANK #1		BANK #2		BANK #3		BANK #4		BANK #5	
CH	FREQUENCY	CH	FREQUENCY	CH	FREQUENCY	CH	FREQUENCY	CH	FREQUENCY
001		041		081		121		161	
002		042		082		122		162	
003		043		083		123		163	
004		044		084		124		164	
005		045		085		125		165	
006		046		086		126		166	
007		047		087		127		167	
008		048		088		128		168	
009		049		089		129		169	
010		050		090		130		170	
011		051		091		131		171	
012		052		092		132		172	
013		053		093		133		173	
014		054		094		134		174	
015		055		095		135		175	
016		056		096		136		176	
017		057		097		137		177	
018		058		098		138		178	
019		059		099		139		179	
020		060		100		140		180	
021		061		101		141		181	
022		062		102		142		182	
023		063		103		143		183	
024		064		104		144		184	
025		065		105		145		185	
026		066		106		146		186	
027		067		107		147		187	
028		068		108		148		188	
029		069		109		149		189	
030		070		110		150		190	
031		071		111		151		191	
032		072		112		152		192	
033		073		113		153		193	
034		074		114		154		194	
035		075		115		155		195	
036		076		116		156		196	
037		077		117		157		197	
038		078		118		158		198	
039		079		119		159		199	
040		080		120		160		200	

NOTES, COMMENTS, CALCULATIONS, ETC:

Figure 2-1b
MASTER PROGRAM PLAN FOR 400-CH SCANNERS

PAGE____OF____
PROGRAM OR BLOCK No:____
DATE:_____

CH	BANK #6 FREQUENCY	CH	BANK #7 FREQUENCY	CH	BANK #8 FREQUENCY	CH	BANK #9 FREQUENCY	CH	BANK #10 FREQUENCY
201		241		281		321		361	
202		242		282		322		362	
203		243		283		323		363	
204		244		284		324		364	
205		245		285		325		365	
206		246		286		326		366	
207		247		287		327		367	
208		248		288		328		368	
209		249		289		329		369	
210		250		290		330		370	
211		251		291		331		371	
212		252		292		332		372	
213		253		293		333		373	
214		254		294		334		374	
215		255		295		335		375	
216		256		296		336		376	
217		257		297		337		377	
218		258		298		338		378	
219		259		299		339		379	
220		260		300		340		380	
221		261		301		341		381	
222		262		302		342		382	
223		263		303		343		383	
224		264		304		344		384	
225		265		305		345		385	
226		266		306		346		386	
227		267		307		347		387	
228		268		308		348		388	
229		269		309		349		389	
230		270		310		350		390	
231		271		311		351		391	
232		272		312		352		392	
233		273		313		353		393	
234		274		314		354		394	
235		275		315		355		395	
236		276		316		356		396	
237		277		317		357		397	
238		278		318		358		398	
239		279		319		359		399	
240		280		320		360		400	

NOTES, COMMENTS, CALCULATIONS, ETC:

banks allocated to things such as VHF or UHF aeronautical, or VHF maritime, ham repeaters, or forestry conservation, or railroads, or other special areas that many monitors especially enjoy. The listing in Table 2-10 is merely a generalized sample. Obviously, each individual would want to customize the programs for their own unique needs and interests. The idea is to create a logical scheme for programming your scanner. The more banks and channels your scanner has, the greater the necessity for a systematic approach to programming the unit. If you have a scanner with 10, or 16, or 20 channels, then relax, there's not much to do. And, if that doesn't beat all, what if you do the 6,400 channel memory modification given elsewhere in this book? Talk about work!

Now think about this: If the scanner has 16 or more channels, there's a good chance that after a short time you'll probably forget what some (or most) of the channels are that you've programmed in. The probability of this increases exponentially with the more thannels your scanner has. Human memory is inadequate to effectively meet the challenge of managing the contents and distribution of 50, 100, 300, 400 or more channels. Chances are that most people with 300 and 400 channel scanners have unwittingly programmed the same frequency into at least two (or more) channel positions. Enjoyment can turn into drudgery and loss of interest if you don't have a system to rescue you from being swallowed up by having to manage more channels than your own memory can juggle at one time.

Here's another point of interest. Rehardless of the capabilities of your scanner, be it 10 or 400 channels, what do you do next when you fill up all the channels? This is easy to do since there are some 500 to 2,000 channels of potential interest in most regions of the nation. This can be a serious problem because no matter how many or how few frequencies are of interest to you, there will always be one more to pop up sooner or later. If your scanner's memory banks are topped off right from the start, then there will no room in which to program additional frequencies you want to monitor. You should make it a point to leave a few blank or empty channels in each memory bank of your scanner. This will allow room for orderly addition of new frequencies as they become known to you. This is one of the greatest values of the exciting 6,400 channel memory modification featured in Chapter 4 of this book. You might never use all 6,400 channels, but you can eventually fill up several thousand and still rest assured that you have an abundance of spares available for more.

The work of organizing your program can be fun and even educational with the right approach. Several blank forms are included here for you to adapt to the requirements of your own situation. Figures 2-1a and 2-1b comprise a **Master Programming Plan for 400-Channel Scanners**. Combined, they contain spaces for for programmed frequencies in up to ten banks of 40 channels each for a total of 400 channels. If you have 200 or less channels, you'll need only one chart. If you do the 6,400 channel memory modification, you might need sixteen double-charts!

Figure 2-2 is a **40-Channel Bank Memory Detail** form to help you focus in on the detail of each channel. Use it to record detailed information about the frequencies you've proframmed. There are spaces for channel-number, frequency, class of radio service, type of user, location, callsign, and descriptive comments. Also, there's a column entitled "Block," which should be used if you do the 6,400 channel memory modification where you'll have up to sixteen blocks of 400 channels per block. If you have "only" 400 channels, you'd want ten such sheets to complete your documentation. If you have 6,400 channels, then you'll need as many as 160 of these sheets. Awesome, isn't it? Well, think how you'd ever find anything if you didn't have any way of keeping track of what was programmed in your scanner!

Figure 2-3 is a **6,400 Scanner Channel Memory Overview** and is mostly for use with those scanners having thousands of channels installed, but it's useful for even 50 to 100 channel scanners. It's simply an overview of the general plan or strategy you have selected for programming your scanner. Use the "400 Ch BLOCK" column to record

Figure 2-2
4Ø-CHANNEL BANK MEMORY DETAIL

BLOCK #:_____ BANK #:_____ DATE:_____ PAGE ___ OF ___

BLOCK	CHAN	FREQUENCY	CLASS	TYPE	LOCATION	CALL	COMMENTS

NOTES & COMMENTS:

Date:_____   Figure 2-3   Page___of___

6,400 SCANNER CHANNEL MEMORY OVERVIEW

```
400 Ch      40 Ch
BLOCK       BANK         DESCRIPTION

_____ >------------->|_____
         1: 001-040      |_____
         2: 041-080 ---- |_____
         3: 081-120      |_____
         4: 121-160 ---- |_____
         5: 161-200      |_____
         6: 201-240 ---- |_____
         7: 241-280      |_____
         8: 281-320 ---- |_____
         9: 321-360      |_____
        10: 361-400 ---- |_____


_____ >------------->|_____
         1: 001-040      |_____
         2: 041-080 ---- |_____
         3: 081-120      |_____
         4: 121-160 ---- |_____
         5: 161-200      |_____
         6: 201-240 ---- |_____
         7: 241-280      |_____
         8: 281-320 ---- |_____
         9: 321-360      |_____
        10: 361-400 ---- |_____


_____ >------------->|_____
         1: 001-040      |_____
         2: 041-080 ---- |_____
         3: 081-120      |_____
         4: 121-160 ---- |_____
         5: 161-200      |_____
         6: 201-240 ---- |_____
         7: 241-280      |_____
         8: 281-320 ---- |_____
         9: 321-360      |_____
        10: 361-400 ---- |_____


_____ >------------->|_____
         1: 001-040      |_____
         2: 041-080 ---- |_____
         3: 081-120      |_____
         4: 121-160 ---- |_____
         5: 161-200      |_____
         6: 201-240 ---- |_____
         7: 241-280      |_____
         8: 281-320 ---- |_____
         9: 321-360      |_____
        10: 361-400 ---- |_____
========================================================
```

the specific block of channels to be described. The "40-Ch BANK" column gives spaces for Banks 1 through 10 and the designated channels per bank. Use the "Description" column to key or title the contents of each major 400-channel block and the minor 40-channel banks.

These three forms are more than adequate to make some sense out of an otherwise unmanageable overkill of information. Adapt them to your purposes and needs. Create new forms and procedures if you like. Then, if you hit upon something that works better, write and let me know about your system. New ideas are certainly welcome and needed in this area. Send your thoughts to me in care of Commtronics Engineering, P.O. Box 262478, San Diego, CA 92126.

OK, I realize that there hasn't yet been any discussion here on the use of computers in conjuction with managing the frequencies in a scanner. I have used a computer and a data base manager software program to manage my frequency and channel data for several years and I'd be lost without it now. Understandably, most scannists probably don't have routine use of a computer, so the procedures given here are geared for the "manual approach." What might surprise you, though, is that the same three forms I've provided (Figures 2-1a/b, 2-2, and 2-3) are the same basic format by which my computer does the work.

Using a Computer. If you have a computer and want to try it out for "Scanner Frequency and Channel Management," I suggest for software that you use a "spreadsheet" for Figure 2-1a/b; a "data base manager" for Figure 2-2; and a "word processor" for Figure 2-3. Because there are so many different computers and such a variety of software, I couldn't possibly make a recommendation as to which is "the best," but for scanner organization programs, nothing very sophisticated is necessary. I use a turbo-modified **Apple IIe** and **Appleworks 3.0** integrated software which contains interactive spreadsheet, data base manager and word processor to manage some 2,500 frequencies active in the San Diego area. My Apple computer has 2.2 megabytes of RAM; warp-speed accelerator; 30-megabyte hard disk drive and two each of 3.5" and 5.25" floppy disk drives; a system which is a lot more sophisticated than actually needed for this purpose. Of course, my computer is used for lots of other things as well. My "Master Frequency and Channel Data Base" is pretty much in the same format and layout as Figure 2-2.

Scanner Operations

If you've paid attention to all of the organizational tips I have given, there isn't much to say about actually using your scanner. That's the fun part which is more-or-less dependent upon how well you've organized things. There isn't anything mystical or magical about how to push a **scan** or **search** button and let the scanner do all the work for you, is there? Still, there might be a trick or two I can pass along.

Instant Weather Channel Access. The NOAA's 162 MHz weather service broadcasts are useful, but if you program one of these stations into your scanner, the scanning action will stop as the unit locks up on the continuous carrier. If your scanner offers a "Priority Channel" feature (usually in channel position 1), you can program your locally active NOAA frequency there and then lock that frequency out with the "lockout" button. While you're performing your regular monitoring activities, the scanner will bypass and totally ignore this frequency. However, when you want to instantly hear the weather, all you need do is press the button that activates the "priority channel." The scanner will stop whatever it is doing and hop right over to the NOAA frequency. When you've finished getting the weather information, deactivate the "priority channel" feature and the unit will resume its regular scanning activities. Pilots might wish to consider this same idea, but substituting the ATIS at their local airport for the NOAA frequency.

Receiving Out-of-Band Frequencies. Here's an old trick for receiving frequencies out of the range that your scanner will allow. First of all, this **image frequency**

method is old hat, but will work <u>only</u> with "dual conversion" scanners that use a 2nd I.F. frequency in the 10 MHz range. This trick will <u>not</u> work on "triple conversion" scanners such as the Realistic PRO-2004/2005, but then these scanners will receive almost everything, anyway, so tricks aren't necessary.

As an example, some units don't have the ability to tune the 406 to 420 MHz federal government band. If your scanner was designed to receive UHF as low as 430 MHz, this method will let you receive almost anything in this rather interesting band. The idea is to program your receiver to the "image frequency" of the desired frequency containing the communications you wish to monitor. The image frequency is actually two different frequencies consisting of the desired frequency plus and minus two times the 2nd I.F. frequency, either of which (plus or minus) can be tried. First, you have to determine the 2nd I.F. frequency of your scanner. The very early Bearcat 250 models had a 10.85 MHz 2nd I.F., but later 250's, as well as other Bearcats (including current Uniden models) use 10.8 MHz. Realistic models (except PRO-2004/2005, as mentioned) use 10.7 MHz. Now, here is the formula to use:

Freq. to program in = Desired Frequency $\pm$ (2 x 2nd I.F. Frequency)

Note: $\pm$ means plus <u>and</u> minus, yielding two different image frequencies that can be programmed into your scanner to receive out-of-band signals-- assuming that your scanner is capable of tuning the images.

Example: Suppose you wanted to receive 417.550 MHz on your Realistic PRO-2002 (which has a 10.7 MHz 2nd I.F.):

Freq. to program in your scanner = 417.550 $\pm$ (2 x 10.7)

= 417.550 $\pm$ 21.4

= 417.550 + 21.4

Frequency 1 = 438.950 MHz, or:

= 417.550 - 21.4

Frequency 2 = 396.150 MHz

It's unlikely that your receiver will receive 396.150 MHz, so you'll program in 438.950 MHz, but either would work equally well if your scanner has the capability. Now, when any signals are present on 417.550 MHz, your scanner will pick them up on 438.950 MHz virtually as clear and strong as if the scanner were actually designed to receive 417.550 MHz. There will be some degradation of signal strength and quality, but not much. This idea will work in all VHF, UHF, UHF-T, and even the 800 MHz bands.

Monitoring Quiet & Seldom-Used Frequencies. Who has the patience to hover over a scanner while quiet, seldom-used frequencies are being "watched" for their occasional flurries of activity, or for activity in the middle of the night, or while you're not at home? Are you going to call in "sick" to say you're waiting for activity on some frequency? Still, you can be certain not to miss any of the action. Here's how.

What you need is a cassette tape recorder with a VOX (voice actuated) function or an external device called a "Tape Recorder Switch." The principle of either is simply that the tape recorder sits there in "pause" mode and does nothing until a signal is received. At that instant, the sound triggers the tape recorder, which records the audio for as long as sound is present. When the sound goes away, the recorder returns to the "pause" mode again.

With an arrangement like that, you can set your scanner to a desired frequency and go to bed, to work, to school, to a movie, or wherever. Come back hours later, and anything that transpired on that channel will have been recorded without any lengthy dead periods between transmissions.

A nifty "Tape Recorder Switch" circuit and detailed instructions are given in this

Tape recorders are handy scanning accessories. First you have to devise a way to get them to record only when transmissions are taking place, then you've got to improve the audio quality that your scanner feeds into a recorder. See text.

book in the chapter on modifications, but if this doesn't interest you, Radio Shack and other companies carry a line of cassette recorders with the VOX function. To work it that way, all you have to do is connect an audio cable between the "headphone" jack (or the "tape rec" jack, if one exists) on your scanner to the "mic" input jack on the recorder. Set the squelch on the scanner. If you are hooked to the "tape rec" output jack of the scanner, it may not make any difference to the recording where you have the scanner's volume control set. If you are using the "headphone" or "earphone" jack, the scanner's volume control setting will have to be turned up to a reasonable level-- best thing to do is try a sample recording on a busy channel and then play it back before you walk off and leave the equipment to itself. Then, put the tape recorder in the "record" mode with the VOX function "on." If the VOX level is adjustable, you may wish to experiment with different settings to assure proper performance when signals come through.

Note: Sometimes, even with the squelch set, the scanner will pass some noise that you can't hear but which will nevertheless be detected by the tape recorder's sensitive VOX circuits. If this happens, the recorder will not pause, even when the squelch is set. If you can't find a suitable setting for the tape recorder's VOX function to overcome this, your only alternative may be to forget the VOX idea and use the external "Tape Recorder Switch" described in this book. However...

...Yet another approach to recording signals without wasting a lot of tape during silent periods is by using an external device sold commercially under the name **Nitelogger** which does essentially the same thing. One cable goes into this device from the headphone jack of the scanner and two cables come out of the device for connection to the recorder, one for the microphone input and the other for the remote jack. Check catalogs and ads in scanner publications for this item.

See MOD-23 in the Modifications Chapter for a low cost retrofit kit that automates the memory storage of new frequencies found in "search" mode. This kit is for the Realistic PRO-2004/2005 and is very useful.

Grounding for Safety & Performance. The metal chassis of all electronic equipment, especially radios, should be connected to a good "true earth ground" for maximum safety and performance. Good grounding practices can save valuable equipment from the effects of lightning and faults in the electrical system. The AC ground as the third wire in a three-wire electrical system is good enough for most safety purposes, but this AC ground is rarely a good RF ground.

An ideal grounding system for a monitoring post will consist of a short run of #6 solid copper wire laid out inconspicuously behind station equipment. The #6 wire will be routed via the most direct path to two or more copper ground rods driven into the earth at least 8-ft. deep. The length of the ground wire between the station and the rods ideally will be less than 8-ft., though as short as possible is even better. Use smaller interconnecting wires, even alligator clips, to connect the metal chassis of your radio equipment to the #6 copper wire hidden behind the equipment. Ground wires should never have any coils, kinks, loops, or sharp bends.

Copper sheathed ground rods can be very expensive and there are better alternatives. Ten foot sections of copper water pipe, 3/4" to 1" diameter, make superb ground rods, provided that the soil consistency will permit them to be driven into the ground at least 8 ft. Drive in at least two, and preferably three or more ground rods, spaced one to five feet apart. Interconnect them together with #6 solid copper wire by soldering or with special grounding clamps available at electrical supply firms. Connect the #6 solid copper wire from the station equipment to the ground rod closest to the station. Counterpoises are not necessary nor even suggested for VHF/UHF radio equipment. The important thing is a direct, short connection from the metal chassis of all your radio equipment to "true earth ground." Safety will be enhanced and actual performance of the equipment sometimes can be notably improved with a good ground.

<u>Never</u> introduce chemicals or salt to your grounding system. The initial slight advantage will soon turn to a liability as the grounding rods react to salt and chemicals by corroding. Corrosion is a great insulator. If nothing else, at least run #14-#18 solid copper wire from the metal chassis of your scanner to the screw that holds the AC wall outlet cover place to the receptacle box. The metal support structure of AC receptacles, and consequently the screw, are usually AC grounded, which is better than no ground at all.

Speaking of no ground at all, please, <u>never</u> cut off the grounding prong from the AC connector on any of your equipment. It was put there for your safety.

Chapter 3

Cellular Telephones: Explained

Chapter 3

Cellular Mobile Telephone, Explained

The Electronics Communications Privacy Act (ECPA) of 1986 expressly makes it illegal to monitor cellular mobile telephone conversations. This came about after much furor, consternation and sheer panic from within the cellular industry and its customers. There were fits of anxiety, apoplexy, indignation, whining, and gnashing of teeth when they realized that 800 MHz-band scanners were available to the general public that could be programmed to pick up cellular calls just as easily as NOAA weather forecasts and the local fire dispatcher. It seems that cellular users had always assumed that the handy car phones offered just as much privacy as landline telephones, and the industry had always reassured these people that such was the case.

After considerble bickering and pleading in Washington, the ECPA was passed by Congress and signed into law by President Reagan. So, now it's illegal to listen to cellular phone conversations, even though your scanner came factory-made to tune those frequencies, or you legally modified your scanner, or got a converter that permits tuning the band. The Federal Communications Commission and the Department of Justice have both expressed serious reservations about their ability to effectively enforce the ECPA, even if it were high on their list of priorities-- which it isn't. A short reaction of common sense suggests that the ECPA, as it applies to the monitoring hobby, is largely an unenforceable toothless tiger and, for that matter, an absurdity that at least impresses cellular customers with the fact that there's now a federal law that guarantees them communications privacy.

On the other hand, the ECPA isn't directed solely against monitoring cellular telephone conversations. Fact is, it's actually illegal to intercept radio communications that originate from or terminate in normally private landlines. In fact, that includes most microwave and satellite transmissions and even broadcast studio-transmitter links near 26 MHz and 950 MHz. It's even illegal now to descramble scrambled transmissions. There are exceptions to the ECPA, including ham and CB transmissions, cordless telephone equipment, aircraft and maritime transmissions, broadcasting stations, governmental communications, and several other categories.

Some have suggested that the ECPA might be unconstitutional, but that would be for the Supreme Court to decide for someone who had the resources to pursue the matter there. The bottom line, however, is that the ECPA doesn't make it illegal for persons to manufacture, sell, own or modify equipment that is capable of tuning the frequencies used for cellular operations. It's just that it's illegal to listen!

This chapter explains the cellular concept and how it operates, because it's quite an interesting arrangement that's quite unlike anything else in the VHF/UHF spectrum. It should be of special interest to engineers, technicians, students of communications systems, and those law enforcement authorities whose activities would be enhanced by a better understanding of how the cellular telephone system works.

The Cellular Concept

Two bands of frequencies between 800 MHz and 900 MHz have been allocated by the FCC for cellular telephone service. This portion of the radio spectrum is noted for short range communications, especially with low power radios and low gain or deliberately inefficient antennas. This unusual combination of limitations, however, joins forces to make the cellular concept work. The cellular idea is low power, short range, and frequency reuse.

The heart of the cellular concept is the **cell site**, which is a base station designed to have a range of 3 to 5 miles. An urban cellular system is designed around a number of cell sites ("cells"), each strategically located so that its area of coverage will not seriously overlap an adjacent cell. A small city or metro area might need as few as three to five cell stations for adequate coverage, whereas a large metro complex might require twenty or more cells to cover the area.

The purpose of the cell site is to <u>relay</u> voice communications to and from mobile telephones within a small and restricted area of coverage, and only for those mobiles within its range. That relay depends upon a special format of high speed digital communications that can be monitored and processed by a computer. Therefore every cell site station has a band of frequencies assigned to it, one or more of which are for digital (non-voice) control circuits and the rest are for voice communications. Each frequency at a cell site has its own transceiver, with tramsmitters operating between 870 MHz to 890 MHz (869 and 894 MHz in some metro areas with expanded channel service). Mobile units transmit 45 MHz lower, 825 to 845 MHz (in some metro areas, 824 to 849 MHz). Each cell site has up to forty-five voice channel transceivers (frequencies), one digital control channel transceiver, and one "locating" receiver. All transceivers, mobile and cell cite, are PLL synthesized and their frequencies can be changed at any time upon command by the system computer.

The mobile unit, when it is initially turned on, automatically scans the digital control channels, and it might receive signals from several cells. The mobile unit will determine which one is strongest and then locks onto that signal-- presumably and usually it is the cell closest to the mobile unit. As the mobile unit travels, sooner or later that particular control signal will fall below an acceptable level at which time the mobile unit again scans the control channels seeking a stronger signal. If it finds one, it locks on to the strongest. If it finds nothing, the mobile unit then goes into its "No Service" or "Out of Service" mode and won't be able to used again until it again comes within range of another cell. Normally, the cells all overlap so that service isn't disrupted within a given region. A major aspect of the mobile unit is its "slave" status, which makes it respond to and do whatever the system computer directs.

In addition to the standard telephone number for call-in purposes, each mobile telephone has its own unique **S**ecurity **I**dentification **N**umber (SIN), which is built into the unit at the factory, and no two SIN's are identical. The SIN can never be changed or removed. This allows the system computer to recognize when any given mobile is active and permits digital communication with the mobile. All cell cites in a given system are connected via landline to the central computer site, usually referred to as the **M**obile **T**elephone **S**witching **O**ffice (MTSO). The MTSO controls the entire system, remote cell sites and mobiles in communication with those sites. Here is what happens within one second when a mobile user presses the "Send" button after dialing in the desired phone number to be called:

1. The phone number being called is transmitted along with the mobile's own telephone number and its SIN number to the nearest cell site.

2. The cell site receives the incoming information from the mobile and automatically feeds it to a landline that is connected to the MTSO computer at the central office of the cellular service supplier.

3. The MTSO checks the mobile's telephone and SIN numbers for errors and

validity. This not only deters theft in the long run, in the short term it will quickly spot a mobile subscriber whose service has been cut off for failure to pay his bills, and will also spot mobile subscribers from other cellular service suppliers whose calls may be refused.

4. Assuming there's no problem, the MTSO selects an empty voice channel at the cell site and sends a digital signal to the mobile unit via the cell site to switch to the desired channel.

5. The MTSO places the call and connects the landline phone to the selected frequency at the cell site.

6. The mobile phone automatically switches to the assigned voice channel for the call.

When a party calls a cellular mobile, the MTSO might not know which cell the mobile is in at the time, so the MTSO sends out a digital "page" signal for the one mobile over all active control channels at all cell sites, then:

1. The mobile, if it is active, receives the page on a control channel from a given cell cite and automatically responds back with its identification numbers.

2. The cell site passes that response via landline to the MTSO which is then aware that the mobile is available to receive a call, and knows in which cell site it is located. The MTSO selects an unused voice channel at that cell site and sends a message over the control channel to direct the mobile to switch to the desired channel.

3. The MTSO switches the landline call to the selected radio voice channel.

4. The mobile automatically switches to the selected voice channel.

5. The mobile unit rings to let the subscriber know a call is being attempted.

6. If the mobile subscriber doesn't answer the call within a short time, the MTSO switches the incoming call to a recording that advises there is no answer in the mobile unit.

As a mobile travels, it may pass from one cell to another. As this happens, the signal from the mobile will fade at the first cell site and increase at another cell site. All cells have a "locating receiver" that monitors and measures the signal strengths from all mobiles. When the signal at the first cell site falls to an unacceptable level, the following sequence happens:

1. The MTSO compares all the signal reports from the surrounding cell sites and selects the site that is getting the best signal.

2. The MTSO selects an unused voice channel at the new cell site and sends a digital control signal over the current voice channel being used in order to command the mobile unit to switch to the new channel.

3. The MTSO reconnects the circuits to the equipment at the new cell site and the conversation continues undisturbed and without interruption.

This sequence is called a **hand off** and requires less than a third of a second to complete. The effects are not very noticeable to the persons conversing.

Cellular Frequency Allocations

In systems using the 870 MHz to 890 MHz frequency band for their cell sites, there are 666 channels spaced at 30 kHz intervals. In expanded frequency systems used in some metro areas (cell sites transmitting on 869 to 894 MHz), there are 832 channels spaced at 30 kHz intervals. All cell site/mobile frequency pairs are exactly 45.000 MHz apart, with the mobile units on the lower frequency of the pair (i.e., cell cite on 873.810 MHz would be paired with mobiles on 828.810 MHz).

By law, two companies in any given locality are permitted to operate cellular mobile telephone services. One service provider is always a telephone company and the other provider may be anything <u>except</u> a telephone company. In cellular service jargon, these two types of companies are referred to as **wireline** and **non-wireline** providers. Both companies provide essentially the same services; the only differences to the customer perhaps being the cost and qualitative aspects of the service. In other words, one company might charge more or less than the other, one company's system might have fewer dead spots, one company might offer call forwarding or other extra optional services. But, mostly they are pretty much the same, except for the frequencies on which they operate. The two companies do not share the same frequencies.

The wireline provider's cell cites in a system in the 870 to 890 MHz band would occupy the 880 to 890 MHz portion, while the non-wireline provider's cells transmit in the 870 to 880 MHz portion of the band. Remember, the mobile units associated with these systems operate frequencies paired exactly 45.000 MHz lower. This offset permits duplex (simultaneous) two-way conversations as on a landline telephone.

Table 3-1 shows the specific bands allocated to cells (bases) and mobiles of the two types of service providers operating in the 870 to 890 MHz format system.

Table 3-1

<u>CELLULAR BAND FREQUENCY ALLOCATIONS</u>

Wireline (telephone company) cell sites (bases):   880.020 - 889.980 Mhz
Wireline (telephone company) mobiles (car phones):   835.020 - 844.980 MHz

Non-wireline company cell site (bases):   870.030 - 879.990 MHz
Non-wireline company mobiles (car phones):   825.030 - 834.990 MHz

Since cellular systems are computer controlled and operated, the digital data channels are always going full blast with an annoying buzzsaw sound. These control frequencies are shown in Table 3-2.

Table 3-2

<u>CELLULAR MOBILE TELEPHONE COMPUTER CONTROL FREQUENCIES</u>

Wireline (telephone company) cell site (bases):   880.020 - 880.620 Mhz
Wireline (telephone company) mobiles (car phones):   835.020 - 835.620 MHz

Non-wireline company cell site (bases):   879.390 - 879.990 MHz
Non-wireline company mobiles (car phones):   834.390 - 834.990 MHz

With 30 kHz channel-spacing, in a typical 870 to 880 MHz, or 880 to 890 MHz system, there are twenty-one computer control channels and 312 channels for voice, for a total of 333 channels for each service provider. This, then, breaks down into what might be considered several voice bands for cell sites and mobiles:

Band #1 870.030 to 879.360 MHz (Non-wireline cell sites)
Band #2 880.650 to 889.980 MHz (Wireline cell sites)
Band #3 835.650 to 844.980 MHz (Non-wireline mobiles)
Band #4 825.030 to 834.360 MHz (Wireline mobiles)

The bases (cell cites) use more power than the mobile units, and have antenna systems that are higher and more formidible than the mobile units. As a result, the cell sites present strong signals. Moreover, in almost all instances, the cell sites transmit both sides of all conversations inasmuch as they repeat the received signals from the mobile phones with which they are in communication.

You might wish to refer to Tables 3-3 and 3-4 which depict the unique frequency layout for up to seven cells. This is a complete cellular system frequency layout plan for wireline and non-wireline systems. Visualize a system this way: In order to avoid adjacent (side-by-side) cells from having the same frequencies to interfere with one another, seven cells are required; one at the center and six more surrounding the center cell. There is no particular pattern as to how Cells "A" through "G" have to be laid out. That is, Cell "D" can just as readily be a center cell with the others circling it, as could any other combination. In a metro system consisting of many cells, there isn't any such thing as a "center" cell, because every cell is, in effect, a "center cell" with respect six others which surround it.

Generally speaking, two cells can (and do) operate on the same frequencies when they are separated by <u>at least one different cell</u>. Actually, the seven cell system unit as depicted in Figure 3-1 is used over and over. Two or even more adjacent cells on different frequencies are located between any two cells on the same frequencies. The cellular concept thus takes advantage of low powered, short range 800 MHz propagation to reuse the same frequencies at several different cell sites in a large metro region. If this weren't possible, then only 312 simultaneous conversations could take place at any one time, as it is thousands of simultaneous conversations could be accomodated within a large cellular system, thanks to frequency reuse.

Another factor here is the unique side effect of Frequency Modulation (FM) where an FM receiver <u>exclusively</u> "hears" the stronger of two signals presented to it on the same frequency. So when cells on the same frequency are separated by one or more cells, even though a mobile might be positioned to detect signals from either, it actually will accept only the strongest one. The odds are very slim of the mobile being located precisely where the two signals are exactly equal. But even in that case, the odds against interference are improved even more because chances are virtually certain that the mobile would be under the control of a stronger third cell site signal on a different frequency.

Not only do adjacent cells not use the same frequencies, but no two cells use adjacent frequencies. For example, a given cell (Cell "D") that transmits on 880.950 MHz will not transmit on 880.980 MHz nor on 880.920 MHz. Likewise, mobiles within any given cell will not transmit on adjacent frequencies. This arrangement prevents adjacent channel interference in receivers located at cell sites and mobile units. FM receivers are not very selective to begin with, and the use of adjacent channels would cause interference within a cell. The scheme depicted in Tables 3-3 and 3-4 was created to minimize the chances of adjacent channel interference throughout the entire cellular system. Note that each cell is allocated 47 or 48 frequencies, with a spacing of 210 kHz (seven channels) between each assigned frequency. In that manner, adjacent frequencies are not used in the same or adjacent cells.

Discussion of Figure 3-1. Figure 3-1 illustrates the concept of a very large cellular mobile telephone system. Cities and metro complexes are rarely symmetrical due to geographical and other considerations, so Figure 3-1 is elongated to simulate the configuration of a realistic cellular network. Cities tend to grow along railroads, rivers, and major highways, so the cellular system here is designed accordingly. Most are not this large, with the typical system consisting of three to seven cells. Small communities might even be served with a single cell, while metro areas like Los Angeles and New York City might consist of a number of interconnected systems fanned out to form a huge network. Frankly, size doesn't matter, because of low power, short range, and frequency reuse. The potential size of a cellular system is unlimited, so let's use Figure 3-1 to discuss how a "typical" system is structured:

Figure 3-1
Typical Cellular System Layout

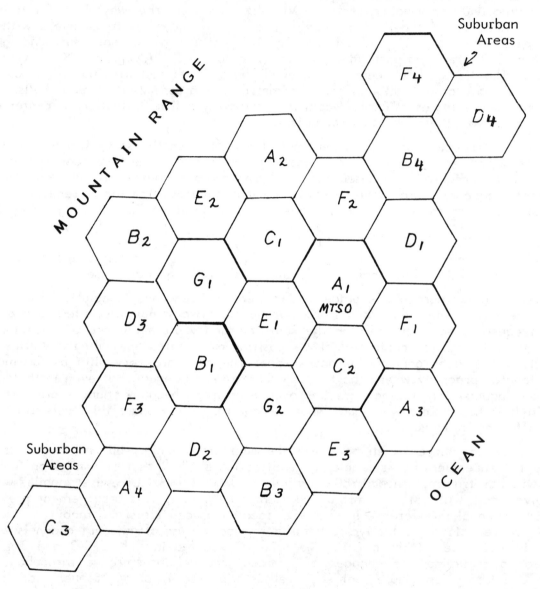

Notes - See Text

1. Cells of the same letter operate on same frequency groups. See Tables 3-3 & 3-4.
2. Numerical designator distinguishes cells of the same letter/frequency group-- otherwise there is no difference.
3. Two companies are permitted to operate cellular systems in any given metro area. The two systems will be laid out functionally as shown above, but the physical layouts will be different.

1. A hexagon is used to depict a cell's coverage territory, but the actual coverage wouldn't be that shape; it world be more-or-less circular, depending upon terrain and geography. However, circles don't illustrate the cellular concept as well as hexagons, and that is why hexagons are usually used in diagrams of cellular systems.

2. No two adjacent cell sites use the same frequencies. In other words, two Cell "A's" are never side-by-side, nor two Cell "B's," nor Cell "C's," etc. At least one cell site on different frequencies is always located between two other cell cites that are assigned the same frequencies.

3. No two adjacent cell sites are assigned adjacent frequencies. So, Cells "A" and "B" are never located next to each other. Neither are Cells "A" and "G," or "B" and "C," etc. At least one different cell site is always located between two other cell cell sites that are assigned adjacent frequencies.

Summary: Each cell site is always assigned frequencies that differ by 60 kHz or more from cell sites that are adjacent to it.

This information, while perhaps boring to lay readers, might be very useful or handy to persons such as law enforcement officers performing court-warranted electronic surveillance on cellular conversations of a drug dealer-- inasmuch as DEA and other enforcement officials have long been aware that cellular phones have become heavily used by drug traffickers.

So, let's say that an authorized surveillance is taking place and the suspect is monitored on 880.740 MHz, which is depicted in Table 3-3 under Cell "D." Everything's fine, and the suspect starts to advise his party to meet him at------, and then right at the crucial moment, the suspect's car enters the control of a different cell site, and **presto**, the channel goes dead. Putting the scanner into "Limit Search" mode in an attempt to track the conversation would bring only frustration; might as well have a cup of coffee and call it quits for the night. Chances are that the suspect's resumed conversation will not be encountered. The "Search" mode tracks in a linear, consecutive-frequency order, either higher or lower. If the suspect's conversation should be relocated, it would certainly take a while.

There would, however, be a way of increasing the chances of zeroing back in on the suspect. First, the scanner would have to be programmed with each individual cellular frequency in order by cell sites as depicted in Table 3-3 or 3-4. For such an operation, it would be highly beneficial to be working with a Realistic PRO-2004/2005 that has undergone the 6,400 channel memory modification outlined in this book (MOD-16) so that wireline and non-wireline cell site channels could be programmed.

There wouldn't be any reason to program any of the data-only control channels, but the scanner could be programmed with Channel 1 = 880.650 MHz; Channel 2 = 880.860 MHz; Channel 3 = 881.070 MHz, etc. Channel 40 would have 888.840 MHz, then continuing with Ch. 41 = 889.050 MHz and ending all Cell "A's" programming with Ch. 45 = 889.890. Then, all zero's would be entered into Ch. 45 to 50, with Cell "B" programming as: Ch. 51 = 880.680 MHz; Ch. 52. = 880.890 MHz; through Ch. 95 = 889.920 MHz. All zeros would go into Ch. 95 to 100, and Cell "C" programming would start in Ch. 101 with 880.710 MHz. Get the picture? When completed, the wireline company's 312 voice channel's would have been programmed into the agency's scanner, organized by cell sites and frequency allocations.

This would be particularly useful to the surveillance officer because, as noted earlier, when a mobile unit passes from one cell to another, the new frequency will not be in the old cell's assignment nor will it be an adjacent frequency! Therefore, one could logically eliminate the frequency assignments of three cells from any consideration. So, when the suspect's conversation gets handed off from one cell to another, up to three scan banks that are known not to contain the call are **deselected**. The scanner could then check for the resumed conversation on the remaining sites and probably locate same rather quickly, as in the following example:

Table 3-3

WIRELINE COMPANY CELL SITE TRANSMIT & MOBILE RECEIVE FREQUENCIES

	CELL A	CELL B	CELL C	CELL D	CELL E	CELL F	CELL G
	889.890	889.920	889.950	889.980			
	889.680	889.710	889.740	889.770	889.800	889.830	889.860
	889.470	889.500	889.530	889.560	889.590	889.620	889.650
	889.260	889.290	889.320	889.350	889.380	889.410	889.440
	889.050	889.080	889.110	889.140	889.170	889.200	889.230
	888.840	888.870	888.900	888.930	888.960	888.990	889.020
	888.630	888.660	888.690	888.720	888.750	888.780	888.810
	888.420	888.450	888.480	888.510	888.540	888.570	888.600
	888.210	888.240	888.270	888.300	888.330	888.360	888.390
	888.000	888.030	888.060	888.090	888.120	888.150	888.180
	887.790	887.820	887.850	887.880	887.910	887.940	887.970
	887.580	887.610	887.640	887.670	887.700	887.730	887.760
	887.370	887.400	887.430	887.460	887.490	887.520	887.550
	887.160	887.190	887.220	887.250	887.280	887.310	887.340
	886.950	886.980	887.010	887.040	887.070	887.100	887.130
	886.740	886.770	886.800	886.830	886.860	886.890	886.920
	886.530	886.560	886.590	886.620	886.650	886.680	886.710
	886.320	886.350	886.380	886.410	886.440	886.470	886.500
	886.110	886.140	886.170	886.200	886.230	886.260	886.290
	885.900	885.930	885.960	885.990	886.020	886.050	886.080
	885.690	885.720	885.750	885.780	885.810	885.840	885.870
	885.480	885.510	885.540	885.570	885.600	885.630	885.660
Voice	885.270	885.300	885.330	885.360	885.390	885.420	885.450
Channels	885.060	885.090	885.120	885.150	885.180	885.210	885.240
	884.850	884.880	884.910	884.940	884.970	885.000	885.030
	884.640	884.670	884.700	884.730	884.760	884.790	884.820
	884.430	884.460	884.490	884.520	884.550	884.580	884.610
	884.220	884.250	884.280	884.310	884.340	884.370	884.400
	884.010	884.040	884.070	884.100	884.130	884.160	884.190
	883.800	883.830	883.860	883.890	883.920	883.950	883.980
	883.590	883.620	883.650	883.680	883.710	883.740	883.770
	883.380	883.410	883.440	883.470	883.500	883.530	883.560
	883.170	883.200	883.230	883.260	883.290	883.320	883.350
	882.960	882.990	883.020	883.050	883.080	883.110	883.140
	882.750	882.780	882.810	882.840	882.870	882.900	882.930
	882.540	882.570	882.600	882.630	882.660	882.690	882.720
	882.330	882.360	882.390	882.420	882.450	882.480	882.510
	882.120	882.150	882.180	882.210	882.240	882.270	882.300
	881.910	881.940	881.970	882.000	882.030	882.060	882.090
	881.700	881.730	881.760	881.790	881.820	881.850	881.880
	881.490	881.520	881.550	881.580	881.610	881.640	881.670
	881.280	881.310	881.340	881.370	881.400	881.430	881.460
	881.070	881.100	881.130	881.160	881.190	881.220	881.250
	880.860	880.890	880.920	880.950	880.980	881.010	881.040
	880.650	880.680	880.710	880.740	880.770	880.800	880.830
Digital	880.440	880.470	880.500	880.530	880.560	880.590	880.620
Control	880.230	880.260	880.290	880.320	880.350	880.380	880.410
Channels	880.020	880.050	880.080	880.110	880.140	880.170	880.200

Table 3-4

NON-WIRELINE COMPANY CELL SITE TRANSMIT & MOBILE RECEIVE FREQUENCIES

	CELL A	CELL B	CELL C	CELL D	CELL E	CELL F	CELL G
Digital Control Channels	879.900	879.930	879.960	879.990			
	879.690	879.720	879.750	879.780	879.810	879.840	879.870
	879.480	879.510	879.540	879.570	879.600	879.630	879.660
					879.390	879.420	879.450
Voice Channels	879.270	879.300	879.330	879.360			
	879.060	879.090	879.120	879.150	879.180	879.210	879.240
	878.850	878.880	878.910	878.940	878.970	879.000	879.030
	878.640	878.670	878.700	878.730	878.760	878.790	878.820
	878.430	878.460	878.490	878.520	878.550	878.580	878.610
	878.220	878.250	878.280	878.310	878.340	878.370	878.400
	878.010	878.040	878.070	878.100	878.130	878.160	878.190
	877.800	877.830	877.860	877.890	877.920	877.950	877.980
	877.590	877.620	877.650	877.680	877.710	877.740	877.770
	877.380	877.410	877.440	877.470	877.500	877.530	877.560
	877.170	877.200	877.230	877.260	877.290	877.320	877.350
	876.960	876.990	877.020	877.050	877.080	877.110	877.140
	876.750	876.780	876.810	876.840	876.870	876.900	876.930
	876.540	876.570	876.600	876.630	876.660	876.690	876.720
	876.330	876.360	876.390	876.420	876.450	876.480	876.510
	876.120	876.150	876.180	876.210	876.240	876.270	876.300
	875.910	875.940	875.970	876.000	876.030	876.060	876.090
	875.700	875.730	875.760	875.790	875.820	875.850	875.880
	875.490	875.520	875.550	875.580	875.610	875.640	875.670
	875.280	875.310	875.340	875.370	875.400	875.430	875.460
	875.070	875.100	875.130	875.160	875.190	875.220	875.250
	874.860	874.890	874.920	874.950	874.980	875.010	875.040
	874.650	874.680	874.710	874.740	874.770	874.800	874.830
	874.440	874.470	874.500	874.530	874.560	874.590	874.620
	874.230	874.260	874.290	874.320	874.350	874.380	874.410
	874.020	874.050	874.080	874.110	874.140	874.170	874.200
	873.810	873.840	873.870	873.900	873.930	873.960	873.990
	873.600	873.630	873.660	873.690	873.720	873.750	873.780
	873.390	873.420	873.450	873.480	873.510	873.540	873.570
	873.180	873.210	873.240	873.270	873.300	873.330	873.360
	872.970	873.000	873.030	873.060	873.090	873.120	873.150
	872.760	872.790	872.820	872.850	872.880	872.910	872.940
	872.550	872.580	872.610	872.640	872.670	872.700	872.730
	872.340	872.370	872.400	872.430	872.460	872.490	872.520
	872.130	872.160	872.190	872.220	872.250	872.280	872.310
	871.920	871.950	871.980	872.010	872.040	872.070	872.100
	871.710	871.740	871.770	871.800	871.830	871.860	871.890
	871.500	871.530	871.560	871.590	871.620	871.650	871.680
	871.290	871.320	871.350	871.380	871.410	871.440	871.470
	871.080	871.110	871.140	871.170	871.200	871.230	871.260
	870.870	870.900	870.930	870.960	870.990	871.020	871.050
	870.660	870.690	870.720	870.750	870.780	870.810	870.840
	870.450	870.480	870.510	870.540	870.570	870.600	870.630
	870.240	870.270	870.300	870.330	870.360	870.390	870.420
	870.030	870.060	870.090	870.120	870.150	870.180	870.210

Suspect is on a frequency in Cell "D" when the call is switched. The officer immediately knows that the new cell will not be "C," "D," or "E," so those are **de**selected and the scanner does not bother with them. The suspect will be on only one of about 180 possible frequencies, which the officer could locate within 30-seconds or less if he knows what to do and can react quickly enough. If he had unsuccessfully used the "search" to look for resumed conversations, there were more than 300 frequencies to check through that way. Note: If the suspect was originally in Cell "A," then Cells "B" and "G" can be eliminated as possibilities. Likewise, if the original call was in Cell "G," then calls from Cells "A" and "F" would be eliminated.

Remember: Cells of the same and/or adjacent frequencies are never physically located next to one another! A judicious law enforcement surveillance expert would use <u>both</u> the "scan banks" and the "search" feature as tools to relocate a handed-off cellular conversation.

Note: Cellular handoffs occur quite rapidly, especially when a mobile goes from one cell through the fringe area of a second and then soon after into a third cell. The two handoffs could take place within seconds, and a search for the first handoff could well be in progress when the second handoff takes place. That's when a cell map of a particular area or system would come in handy. Sometimes they can be obtained from cellular service providers.

Chapter 4

Performance Modifications

Chapter 4

Performance Improvement Modifications for Scanners

Many of the modifications in this First Edition are for the popular Realistic PRO-2004 and PRO-2005 scanners, but please don't conclude that the author is unfairly and exclusively biased in favor of these units or of Radio Shack. Due to its enormous capabilities and the ease with which it may be modified, the PRO-2004/2005 has become the favorite of serious monitoring enthusiasts and, in fact, in that respect it's almost the "only game in town" at present. Well, maybe that's a bit of an overstatement, but it holds many elements of truth. Let me explain.

By the time 1978 had rolled around, I was a technical expert on more than fifty brands and hundreds of models of CB radios. Part of the reason was because there were more than 20-million avid CB'ers in a market that was even wilder and crazier than than the CB'ers themselves. There were thousands of CB technicians and a massive grapevine along which which was passed tremendous amounts of information about the technology of CB. There were several technical trade journals, two-dozen CB hobby magazines, and an unending parade of "how to" books, all directed towards the CB trade and hobby. I was armed with the vast resources of that library; massive hype in overkill proportions from a huge and thriving market; direct support from the manufacturers, and a large customer base that was international in scope. At any given time, in those days, I had thirty to fifty CB radios parked in my shop awaiting repair or modification. I couldn't help but become experienced with fifty different brands and hundreds of models. It was kind of forced upon me, more or less.

I knew which models were the clunkers, and which were the gems. And it didn't hurt that at least three publishers pumped out CB service and repair manuals on a monthly basis so that all but a few of the most obscure no-name cheapie imports were covered. Each volume, costing about $5, contained service and repair info on anywhere from five to twenty radios. What that reference library, there wasn't anything that couldn't be made to work again, or modified to exceed its original capabilities.

By contrast in any given year, only one or two scanners ever came into my shop for repair, and none for modification. There was no such thing as a tech library for scanners and each service manual had to obtained separately from the manufacturer at about ten bucks a hit. This situation hasn't changed much, but one harbinger of possible transition was the introduction of the Realistic PRO-2004 in late 1986.

Since then, a new era has been unfolding. Many PRO-2004's have come across my service bench-- rarely for repair, mostly for modification. Modern scanners are very reliable. If the owner takes reasonable care, the scanner should operate without problems for a very long time. By comparison, many CB sets of the 1970's were cheaply designed, mass produced and marketed, and were very susceptable to failure from their own inherent design and manufacturing faults to outside factors such as antenna problems. Short of lightning strikes, voltage surges, or being left out in the rain, outside influences have little effect on failure rates of scanners. Modern scanners are a product of the digital revolution, their reliability is exceptional thanks to advanced manufacturing techniques, better voltage stabilization, and higher reliability components.

Another thing to consider is the market itself. There are only five scanner manufacturers of any significance-- Realistic (Radio Shack); Uniden/Bearcat; Uniden/Regency; Dynascan Cobra, and AOR. J.I.L. hasn't been in evidence for quite some time and may no longer be active. ICOM and Yaesu are really a step beyond the "scanner" market. So, with five manufacturers and maybe fifty base and handheld models among them, there isn't a wide range from which to choose.

That's not the only problem, because it would be easier to become familiar with fewer brands and models. However fewer manufacturers and models is a sign that this market is obviously smaller than was CB, and has a smaller consumer base. In the late 1970's there were thousands of shops that did nothing else but sell and service CB equipment, antennas, and accessories. There are probably very few (if any) shops that sell nothing but scanners and related products/accessories, although some national mail-order firms specialize in scanner products, often supplemented with ham, CB, video, computer, or telephone products.

So, compared to what CB was in its heyday, the scanner market is relatively small. Whatever size it is, say a million or so avid hobbyists, they don't support or foster an "aftermarket," except perhaps in the area of frequency directories. This is to say, some 90% of scanner owners surface every few years to buy a scanner and perhaps replace their antenna system, and then they vanish until its time to replace the scanner again. Most scanner owners aren't aware that their scanners can be modified and made better-- of course, they might not be interested even if they did know. Scanning just isn't the wild type of hobby that CB was, with its adherents constantly seeking the next station-improving accessory, and the next, and the next...

With only a few exceptions, scanners just aren't serious, high-performance radios. Radio Shack has a couple of top-of-the line rigs to interest the serious monitoring enthusiast. AOR and Regency each offer a couple, and Bearcat has maybe two or three. Cobra produces several models. The majority of what's available, with the exception of the few units I mentioned, are rather simplistic scanners that offer limited frequency coverage, no choice of receiving modes, relatively few memory channels, slow scanning speeds, no search capabilities, no ability to change frequency increments in search mode, and other frills.

Here's the problem in crystal-clear perspective: San Diego, CA is a large metro area with several million inhabitants. Every resource imaginable is in the area, but I couldn't tell you where in the metro area to walk into a store in or around San Diego and buy a new Bearcat scanner! One CB shop I know of might handle a Uniden scanner now and again, but certainly doesn't have the complete line on display or in stock. Another CB shop stocks a few AOR scanners, but they're across town.

This type of thing has caused the person who wants to really get serious about scanner monitoring to take the path of least resistance. That person takes a short trip over to the closest Radio Shack (7,000 locations) and buys a Realistic PRO-2004/2005. That's why there are probably more PRO-2004/2005 scanners in use than all other scanners combined. This at least implies, if not directly points to, why everybody is so interested in mods for these sets, and why much of the information in this book is directed at these scanners.

Here is another factor. Perhaps, because the PRO-2004/2005 is so sophisticated a unit, that it has attracted a large following among technical and hacking people than any other scanner models. There are few other radios to intrigue or interest people who get down to the serious interest business of hacking and modifying. In my case, I simply had lots of opportunities to work with these units in order to build a detailed working knowledge far greater than any other make or model. They just don't come in to the shop for modifications, and it doesn't appear that there is much curiosity in regard to most of the less sophisticated scanners. My extensive correspondence with scanner monitors throughout the word is is almost exclusively concerned with modifying the PRO-2004/2005.

Note also that the two leading scanner hobby publications, Popular Communications and Monitoring Times have both devoted considerable space to modifications for the PRO-2004/2005, with very little similar information regarding other scanners.

So, basically, there's simply lots more information in existence relating to the Realtic PRO-2004/2005 than any other ten scanners combined. If you know of additional mods for the PRO-2004/2005, or of mods for other scanners, we certainly would like to hear about them from you. We can publish only what we know about.

While many of the mods given in this edition are PRO-2004/2005 specific, most can be adapted to other scanners with a little ingenuity and minimal grief. I used to some of these mods years ago on CB rigs, and simply figured out how to apply them to the PRO-2004/2005 because the set lent it self so readily to being modified. Others are more specific, but depending upon the scanner, some modifications might also be readily adapted. The principles of the 6,400 channel modification (MOD-16) are one example, which probably can apply to many other scanners that use 8-bit static random access memory chips. Some have told me that this modification can be easily applied to increase the memory of the ICOM IC-R71A and IC-R7000. So if you see a modification that you would like to apply to your scanner, but aren't sure how to do, try this: send me a <u>complete copy</u> of your radio's Service Manual <u>and a self-addressed, stamped envelope</u> with one loose extra stamp (U.S. stamps only), and I'll make a serious effort to point you in the right direction. That's how I'm committed to your enjoyment of radio. My address is: Bill Cheek, Commtronics Engineering, P.O. Box 262478, San Diego, CA 92126.

Modifying Scanners

Absolutely every thing made or contrived by the mind and hand of humans has the potential for modification or enhancement. In other words, if a human hand made it, then chances are a hundred-to-one that it can be made to work better or differently than what the manufacturer intended. Some things are more readily or easily modified than others, like buildings and cars. In some cases, major "aftermarket" industries arise to meet the needs for modifications and consumer alterations. Examples include home improvement centers, automotive parts stores, audio accessory stores, camera shops, and so on. There's always room for improvement. Scanners are prime candidates.

In many instances, for the short-term, consumers get the products and features an industry decides to give them. In the long-term, supply and demand becomes a factor. The exception is notable when new ideas and products from the leading edge of technology or which are complicated beyond the average hobbyist's ability to comprehend. When that technology is in a constant state of evolution and progress, as with scanners, the hobbyist has almost no input in determining the products that show up on the market. Perhaps if there were an opportunity for hobbyists to have input with scanner manufacturers, we might demand and get considerably different functions and performance specs than are presently available.

Monitoring the VHF/UHF bands isn't an especially old hobby. It began in the late 1940's as an offshoot of the ham and SWL hobbies, and as two way communications began to shift from HF to frequencies above 30 MHz. These were frequencies that, until technological developments took place in WWII, had been thought of as being experimental in nature. The first hobby equipment was homebrewed and converted military surplus, then a few simple tunable 30 to 50 MHz and 152 to 174 MHz receivers appeared. Still, this hobby was being pursued by very few people until the invention of the scanning receiver in 1968. For all reasonable purposes, the hobby nominally dates back to the late 1960's. Since that time, each new year has brought major, new developments in technology. There simply hasn't been time for the old law of supply and demand to apply itself to scanning. We take what we are given as manufacturers scramble to apply each new technology to what they believe will become a demand. But, now the time is ripe for the monitoring enthusiast to apply some of that same technology in order to improve that great scanner that still doesn't do quite everything one would hope for.

Modern scanners are integrated along two distinctly different technologies, analog and digital. Analog technology is getting "old hat," while digital is relatively new. When the two technologies are integrated, as in the case of scanners, the result is almost a third technology. Because of that, until fairly recent times, scanners weren't interesting candidates for modification and retrofit performance upgrades. The technology and required level of effort were just beyond the capabilities of a great many hobbyists. This is contrasted with ham and CB radio, both of which have long supported entire industries dedicated solely at aftermarket and retrofit modification packages.

VHF/UHF monitoring is now causing interested ears to perk up. One reason for this appears to be that a number of VHF/UHF communications professionals (engineers and technicians) as well as computer pros are coming into the monitoring hobby. Some of these people have adapted or applied their valuable skills to refining or improving their scanners. Word of mouth, and mentions in the hobby media, about these refinements have created a stir around the hobby. The result is that we don't have to be satisfied with "stock" equipment. We now know that it can be enhanced, and we know how to bring those improvements about in short order.

Almost any scanner can be improved upon or otherwise have more functions and features added by the "electronics hacker." Only a little comprehension of theory and knowledge of electronic techniques are required for even the most casual hobbyist to perform major improvements to a scanner. There are, of course, definite limitations and certain "impossibilities" that must be confronted and accepted.

Adding New Bands of Frequencies

Without any doubt, the most popular and in-demand modification relates to being able to get the scanner to receive more bands or frequencies than those the manufacturer intended. This is a discussion on the several different areas of your scanner that can impose strict limitations on what might be done in the way of enhanced frequency coverage.

1. The Central Processor Unit (CPU): The CPU in a digital, programmable scanner is actually on a chip. To be sure, it's typically an 8-bit digital computer. As a reference, the Commodore 64, Commodore 128, Apple II series, and the Tandy TRS-80 are all "8-bit computers." Don't be negatively impressed by the term "8-bit" because it doesn't sound like much, there's a lot of processing power in a programmable scanner. No, it won't do all of the things a stand-alone computer can do, but in its own way it can do just as much or more!

The only problem with a CPU is that you're more-or-less stuck with whatever it comes programmed with inside. The CPU contains a permanent program that directs almost every aspect of the scanner. The program is fixed and resides in the CPU's permanent internal memory. It can't be modified or altered. It can't even be erased. It is "burned" in by the chip's manufacturer and that's that. The CPU program starts automatically when the scanner is turned on.

One neat thing about scanner CPU's is that they are very expensive to produce in small quantities, so the manufacturers contract for lots of to be made right from the very start. The CPU is a "custom" type of a chip especially designed and programmed to do the specific things that a manufacturer wants it to do, and maybe even a little bit more. Savvy manufacturers sometimes design their CPU's to do more than might be desired for a particular model year. When the current model year's scanner chassis is designed and built around that CPU, those extra little performance features are just defeated or wired out with a couple of inexpensive, external diodes or or resistors. The new chassis for the following year might then use the same CPU and have all of the original functions from the previous model, plus those neat extras as standard features!

A manufacturer might also use the same CPU in several different current models, which means that the CPU in all models has to be programmed for all ot the features of the top-of-the-line model. For lower cost scanners using this same CPU, the deluxe features might simply be wired out with diodes and resistors. Therefore, the CPU in your scanner could very well possess some latent and repressed capabilities that are practically shouting and screaming to be liberated.

This is pretty much the only way to gain extra bands or frequency coverage. If the capabilities weren't designed there in the first place, then you're stuck with what you've got. Again, the internal programming of the CPU is rigid and inflexible. If you aren't an engineer, technician or otherwise can't easily find your way around the innards of a modern scanner, it's best to leave things alone and wait for experienced hackers to work up and publish the mods for your particular CPU.

2. RF Sections: Preselector, Mixer & VCO: The function of the "preselector" in a scanner is to admit or pass desired frequencies and simultaneously reject or eliminate undesired frequencies. The thing to keep in mind here is that which the manufacturer desires or rejects doesn't necessarily coincide with what you want. Usually, the manufacturer has made the decision for you and designed a preselector that accepts only certain bands of frequencies. This doesn't prevent the manufacturer from using the same CPU chip in several different scanner models, however. That's where modifications begin to get especially tricky. The CPU may be programmed for 800 MHz-band operation, and you could even "liberate" it to restore its blacked-out capabilities there. But, even though you can now program-in 800 MHz frequencies, and they show up on the display, the scanner isn't going to pick up any signals there unless its preselector has been designed to do so. Unlike a CPU, it would be economically ignorant to build an 800 MHz preselector and then "wire it out" or "hide" it in selected models. Case examples here include Realistic PRO-32 and PRO-2002. In either scanner, adding or snipping selected diodes will let the units display and even scan the 800 MHz band, but there is no 800 MHz preselector in those models so reception is impossible. These are only two examples, many other scanners have the same limitation.

Another critical RF circuit in a scanner is known as the "mixer." As with the preselector, the mixer has to be designed to operate on certain bands of frequencies. If the mixer isn't designed for operation on certain frequency bands, then reception will be impossible, despite any inherent capabilities you can restore in the CPU.

Just like the preselector and mixers in a scanner, the the VCO also has to be appropriately designed for reception on specific desired frequency bands. The VCO is driven by the CPU to work with the mixer to tune the desired frequencies. If either the VCO or the mixer are not designed to function on other bands of frequencies, it will not be possible to extend the scanner's frequency coverage.

Summary: In order to successfully modify a scanner to receive bands or frequencies which the manufacturer does not support, first the scanner must contain modifiable circuits in the CPU, preselector, mixer, and VCO. If any one of these is lacking key design elements, then the scanner cannot be modified, in the customary sense, for operation on additional frequencies. By that, I mean that it might possibly be able to be accomplished with extensive redesign and circuit reworking by a skilled engineer or RF technician. This book is for scanner hackers and experimenters, and there are limits as to what the average experimenter can attempt and still have a nice, working, scanner!

Adding Other Features & Enhancements

The most restrictive or limited modification you can do to a scanner is adding frequencies and bands that weren't intended by the manufacturer. We have seen how the CPU's inflexible internal programming pretty much sets the limits as to what may done with adding extra coverage. With the exception of this and perhaps other CPU controlled functions, the sky's pretty much the limit as to the things that might be accomplished with the unit's analog features.

Scan & Search Speed Increases. Increasing scan and search speeds might be a good example to illustrate. In some instances, notably with the Realistic PRO-2004/2005, the digital CPU, indeed, does control the scan and search speeds. However, these and most other scanners also have an analog circuit that can effect or determine speed. Invariably, such circuits are some form of an oscillator connected to drive a port on the CPU. The oscillator generates a special frequency, the reciprocal of which (1 divided by the frequency) is the basic "gate time" that determines scan and search speed. If the "gate time" is decreased by increasing that oscillator frequency, then speed is proportionally increased. It is relatively easy to change the frequency of the "clock oscillator" of a scanner.

Most clock oscillators are crystal-driven, so if you want to remove the stock crystal and install one similar to it but higher in frequency, then the speed will be increased. There are no limits to just how much faster you can make your scanner go, because the CPU and random access memory (RAM) chips have internal speed limits, beyond which things tend to go berserk. No damage will be done, of course; it's just that your scanner won't work properly, if at all. On the other hand, scanners are designed to be as close to 100% reliable as possible, and this means that the designers routinely compromise on critical specifications such as speed in the interests of economics and reliability.

In the instance of speed, designers might find that with a clock oscillator frequency at 10 MHz, 15% of all production scanners will exhibit erratic operation. Given a production run of 5,000 to 25,000 units, the percentage of Quality Assurance rejects would be unacceptable for profit. Therefore, the designers might specify a clock oscillator frequency of 7.37 MHz because that frequency is known to be 99.99% reliable. What this means to _you_ is that your scanner might be capable of 25% to 100% faster speeds and still operate to specifications in all other respects. On the other hand, if you got one of those scanners in the "15%" category, then yours might not work at 10 MHz. You'll never know until you try, and if you do run into problems you can always try lower frequency crystals until you find one that works perfectly. You may still end up with a speed increase!

Adding Signal Strength Meters ("S-Meters"). One factory representative told me that his company didn't design their scanners with S-meters because the cost would be prohibitive. As a former President and CEO of an electronic manufacturing company, I can understand his position, if not fully agree. It's an undeniable fact that for some reason, manufacturers don't build scanners with S-meters. Equally undeniable is the fact that most scannists who are asked for an opinion say that they would appreciate and utilize an S-meter if one were available.

An S-meter is another analog function and isn't difficult to design and install for your scanner. There are several ways to do this, only the general principles of which we will cover here. The easiest way to get a signal meter is to tap the scanner's Automatic Gain Control (AGC) feedback loop and attach a voltmeter to that loop. AGC is developed in the scanner for the purpose of controlling the gain of the RF and IF amplifiers. AGC prevents overdriving from strong signals-- overdriven amplifiers create distortion, intermod, crossmod, and all kinds of objectional and disagreeable interference. The AGC voltage is, however, a small control voltage that is proportional to the strength of the incoming signal. Therefore, a voltmeter attached to the AGC test point will serve as an effective S-meter.

To make this especially useful and electrically sanitary at the same time, a test jack of your choice should be installed on the rear chassis of the scanner, and be internally connected to the AGC test point. If you install a resistor of about 10K-50K between the jack and the AGC test point, then the AGC will be effectively isolated from influences such as short circuits that could happen outside the chassis. Use a digital voltmeter of your choice to connect to the jack so you can display AGC voltages when scanning. Simple, but very practical.

Modern scanners enjoy a profusion of integrated circuits, some of which are practically an entire receiver on a chip-- and some of these chips actually contain an S-meter output function! No, the function isn't used, but that doesn't stop you from experimenting a little, measuring and testing that chip for what it would do if you were to hook up some kind of a meter. Here's an instance where the Service Manual comes in very handy. Most of these manuals depict the chip's internal functions and the pinouts so you can readily see some of the available capability. If not utilized by the manufacturer, well, maybe you can put it to a useful purpose. MOD-12 and MOD-13 in this book show specific S-Meter modifications.

Adding Tape Recorder Function. If your scanner doesn't have a "tape rec" output jack, there's no reason why you can't add one. Most scanners have one, so that might not be your problem. Your problem might be a tape rec output that simply doesn't produce decent quality recordings. There's an easy cure for either situation.

First, understand that modern tape recorders don't require very much in the way of signal in order to work. In fact, one-thousandth of a volt (.001v) of signal is sufficient to produce quality recordings in most tape recorders now. Trouble is that typical stock tape rec output jacks provide up to 600 thousandths (.600v) of a volt which could be far too much for your tape recorder. If tape recordings from your scanner sound nasty and disagreeable, this is almost certainly the cause. The remedy follows, but you must have the Service Manual from your scanner's manufacturer. Follow the same guidelines if doesn't have a tape rec out feature.

Follow the scanner's RF and IF circuits through to the volume control of the set. Locate the specific area called the "detector" which comes right after what is called the "last I.F." stage. The detector is the device or circuit that converts the IF signal into a very weak audio signal. Following the detector are one or more stages of audio preamplification and then the final power amplifier chip that drives the loudspeaker and/or the headphones. With the scanner disconnected from its power source, install a 0.1-uF capacitor at the detector output and before any audio preamp. Install one end (stator) of a 50K trimmer potentiometer to the output side of the capacitor. Then, ground the other end (stator) of the trimmer pot. To the wiper arm (or rotor) of the trimmer pot, run a wire to a newly installed jack on your scanner. Select a jack that will mate with a cable from your tape recorder. Reconnect the power and begin a a sample recording and adjust the trimmer pot for optimum quality recordings with a minimum of noise and distortion. That's it, except that you can refer to MOD-5 in this book (a specific installation for the PRO-2004/2005) to guide you in adapting your scanner's tap rec function.

Automatic Tape Recorder Switch. You can shell out $50+ for an external automatic tape recorder switch to activate your tape recorder when a signal is present, and then stop it when the signal goes off. At any reasonable price, this is a great asset to monitoring because, literally, hours of scanning can be compressed and condensed on to a 45-min. tape with no dead intervals. If you've got $50 to spare, do it. If you'd rather spend only $10 or less on something a lot better and have yourself some fun at the same time, the eyeball MOD-6 for an excellent circuit diagram and technique that can be adapted to just about any scanner.

Improving the Sensitivity of Your Scanner. In general, the only practical way to approach this noble objective is not by internal modification, rather by the addition and use of an external accessory preamplifier. First, be sure that you've read the discussion on RF preamps in the **Tips, Hints & Kinks** chapter. There's a lot you should know about preamplifiers for scanners.

There might be some possibilities, however, for internal modification to result in increased receiver sensitvity. Ordinarilly, the scanner's internal RF amplifier and preselector circuits would be the very last places you'd want to monkey around with because they are extremely critical to low noise and wideband operation of the scanner. On the other hand, you might at least wish to inspect the RF amplifier transistor from the perspective of your unit's Service Manual from the manufacturer. No harm can from from "paperwork." So check out the first transistor device just after the antenna input jack. Manufacturers aren't always in direct touch with the latest, state-of-the-art transistors, and it is entirely possible that your "new" scanner uses a transistor in this crucial spot that is several years old and nowhere near as "hot" as more recently developed devices.

You will need to identify the transistor(s) known as the RF Amplifier, and then obtain a set of specs and performance curves for that specific device. Next, you'll want to review all of the specs and performance curves for the very latest transistors of the same type. In general, you will be critical of three specifications in particular: noise, bandwidth, and gain. When looking for a potential replacement transistor, you must also keep an eye on all the voltage and power ratings of the original transistor as a replacement would need to equal or be superior to them. In all instances, it is **never** worthwhile to replace a transistor based upon voltage ratings alone. The dynamic characteristics of gain, bandwidth and noise are what will assist you in your decision in whether or not to replace the transistor. Improved gain specs, alone, isn't a reason to replace a transistor, because when placed in a circuit, transistors with radically different gain specs can and do all come up with the same circuit gain. The gain a transistor is rated with on a spec sheet is never the same as circuit gain. Equal or better gain **and** bandwidth **with** a lower noise factor is reason to give serious consideration to using, trying, or experimenting with a replacement transistor.

It's never worth your time and trouble to experiment with replacement transistors for the IF stages of the scanner. Gain, noise and bandwidth aren't important in the IF stages. The critical aspect of the scanner's ability to pick up weak signals (sensitivity) is the RF Amplifier transistor, and (to some extent) the First Mixer and its associated circuit. You'd want to be really careful in messing around with Mixers, however, because they are more critical than the RF Amp with regard to stability and balance-- one seemingly minor change, modification, or adjustment and you could very easily upset a delicate equilibrium with results that aren't good. If you know how to compare, select, and work with small signal VHF/UHF devices, then there is significant potential for improvement of your scanner's sensitivity, signal/noise ratio, and immunity to intermod.

This is one of those realms of modification that, for the time being, is best left to the experts and my humble suggestion is that unless you are a qualified RF technician or engineer you'd probably be better off if you don't attempt any transistor replacements. On the other hand, future editions of this book will contain any notable successes of this nature that are reported to me by competent techs and engineers. The mere mechanics of physically removing one transistor and replacing it with another one isn't the problem, because that's something that can be accomplished by most hobbyists who don't have more than two thumbs on each hand. All you'd need to know is, in a given scanner, which transister gets yanked and what it's replaced with. Aye-- that's the question! And that's what we hope to learn from our own experiences and from the information sent in by others. When I spent a lot of time on CB repair work, I had substantial success in finding better replacements for CB (and ham) RF Amplifiers and Mixers, so I'm certain that information will be developing in this area for scanners.

Now a question for the research experts: Has anybody yet found a superior replacement for the 2SC3356 (the RF Amplifier) in the Realistic PRO-2004/2005?

Other Analog Modifications for Scanners. As I previously noted, the sky's the limit for modifying the analog portions of a scanner. Your own imagination is the main limiting factor in this area. For example, one day I was wishing that my PRO-2004 scanner had more of the characteristics and features of a communications receiver. In a flash, it came to me and was easily accomplished. It was as simple as routing a sample of the 455 kHz signal from the 3rd IF to the antenna port of my Yaesu FRG-7700 communications receiver. Now I can use the integrated system package to receive SSB signals; to control adjacent channel interference; and to perform signal analysis not possible with the PRO-2004 alone. See MOD-14 for how to do this.

I'm interested in Survivacom (survival communications) to a serious extent. Survivacom is the adaptation of radio equipment to help individuals, families, and communities after the effects of a mass disaster situation, be it earthquake, severe storm, tornado, flood, volcano, forest fire, environmental or radiation pollution, or act of war. It can be readily assumed after a major disaster, AC power will either be non-existant, sporadic, unreliable, unstable, or otherwise useless/destructive to sensitive digital and communications equipment. It is during such times when access to monitoring the communications in the outside world (that is, everything more than two miles from your location) can offer an enhanced measure of "survivability." Therefore, it follows that designated Survivacom equipment should have operational capabilities for both AC and DC power and would be relatively immune to surges, transients, and spikes caused ny faulty, damaged, or ill-maintained AC power systems. MOD-10 and MOD-11 offer significant contributions to the Survivalcom concept with detailed instructions on how to make your base scanner more transportable, and how to protect your sensitive equipment from the effects of lightning and power surges. Emergency power supplies for scanners are discussed in the chapter on **Tips, Hints & Kinks.**

The modifications in this chapter offer some really hot improvements and feature upgrades for scanner receivers. I hope that your imagination will be tickled into activity and that you will let me know your accomplishments so that they can be included in future editions in this planned series of books. One thing can always be said of human endeavor: it can always be improved. To some, it's adding ketchup to the steak; to others, it's putting custom pin striping on the ZR-1 Corvette. To me, it's taking a stock scanner and making it do more things than anybody had ever dreamed of. Not only will the following mods improve your scanner, but you are invited to provide me with input on improving these improvements. Now, let's get prepared.

Preparation for Scanner Modifications. It doesn't take a skilled engineer or professional technician to perform any of the modifications in this book. It would be most helpful if you are prepared with a little experience at working with a soldering gun and finding your way around a crowded circuit board. But the critical necessities are the proper tools, a great deal of patience, the ability to read/comprehend/follow instructions. I have gone to a great deal of time and trouble to lay out in excrutiating detail the steps of procedure required to obtain the desired results. I have gone into such detail that engineers and technicians will undoubtedly think I've said several times more than I should have-- but then I didn't write this book for them as much as I wrote it for you, the average, everyday, scanner hobbyist who wants to enjoy the hobby to the fullest from every possible vantage point.

So, if you are an experienced type, please bear with us. Many of you will be able to do the work simply from looking at the schematic diagrams and the parts lists, skipping altogether the detailed instructional steps. Others will rely upon every word, and some may wish that there were even more detail. One requirement is imposed right off the top, and I make no bones about the fact that you **must get a Service Manual for your scanner.** No, maybes about this, you **absolutely must** have a Service Manual for your particular unit before you begin poking around the unit's innards. These manuals are a rich and unique source of information of all kinds relative to the modifications outlined in this book. A picture is worth at least ten thousand words, and along the way this book still had to draw a line on how much detail could be given. As much as is given here, the Service Manual written for your specific scanner should be obtainable (usually at minimal cost) from the unit's manufacturer. In case you've misplaced the address of the manufacturer, the information is given in Table 4-1; it was believed correct at publication time.

Table 4-1
SERVICE MANUALS & TECHNICAL SUPPORT FOR SCANNERS

(Realistic Scanners)
Tandy National Parts Center
900 E. Northside Dr.
Fort Worth, TX 76102

(AOR Scanners)
Ace Communications
10707 E. 106Th Street
Indianapolis, IN 46256

(Regency Products)
Regency Electronics, Inc.
7707 Records St
Indianapolis, IN 46226

(Icom Radios)
Icom America, Inc.
1777 Phoenix Pkwy; Suite 201
Atlanta, GA 30349

(Icom Radios)
Icom America, Inc.
2112 - 116Th Ne
Bellview, WA 98004

(Uniden & Bearcat Scanners)
Uniden Corporation
6345 Castleway Court
Indianapolis, IN 46250

(Cobra Scanners)
Dynascan Corporation
6460 W. Cortland
Chicago, IL 60635

(Yaesu Radios)
Yaesu-Musen Usa
17210 Edwards Road
Cerritos, CA 90701

(Icom Radios)
Icom America, Inc.
3150 Premier Dr; Suite 126
Irvine, TX 75063

(Icom Radios)
Icom Canada, Inc.
3071 - #5 Road
Richmond, BC, Canada V6X 2T4

What The Author Will do if You Run Into Trouble

You have to have a Service Manual for your scanner before doing any modifications, otherwise you won't even be able to help yourself should you run into trouble for some unforseen reason. Most authors usually issue a caveat that if you've got questions or problems, you have to figure out the answers for yourself. I am going to go out on a limb with the offer to extend a helping hand if you have trouble worthing through any of the modifications in this book. That offer, however, is qualified as follows:

1. You must have a Service Manual for your scanner.
2. If I don't happen to have a copy of your scanner's Service Manual in my files, you must be prepared to loan me yours (or a photocopy) so that I can understand and and try to solve your problem.
3. You must lucidly communicate the nature and symptoms of your problems in a letter. If you don't use a typewriter, please write clearly and legibly enough so that I won't have trouble trying to make out what you are trying to tell me.
4. You must enclose with your letter a business-sized, self-addressed, stamped (US stamps only) with one loose (unstuck) extra (US only) postage stamp.

5. My offer to assist in resolving any technical problems related to the modifications given herein is limited to written suggestions, discussions, and my personal opinions of potential, probable, and possible remedies-- and only if the foregoing qualifications and limitations are met.

6. Under no circumstances, now or in the future, does the author or publishers of this book assume or accept responsibility for any malfunction of, loss of use of, damage to, legal sanctions related to use of, or any other problems related to your radio equipment or its use, whether or not related to any information given in this book. Your equipment is under your control, and any decisions you make regarding its use or modification are also wholly your responsibility.

7. Under no circumstances whatsoever can there be any telephone discussion of any technical problems or troubleshooting relating to the modifications in this book.

Simply stated, the author will offer written professional opinions and suggestions if you run into trouble with any of these modifications, but you must first have a Service Manual handy, a copy of which I may also require in order to try and help you. I have service manuals on hand for the following Realistic models: PRO-31, -32, -34, -2002, -2004, -2055, -2022.

Direct your technical inquiries to: "Doctor Rigormortis," c/o Operation Assist, Commtronics Engineering, P.O. Box 262478, San Diego, CA 92126-0969. Inquiries sent to any other address will not reach me and therefore receive no reply.

What Else You Need to Perform Modifications. First and foremost is the famous Service Manual. Hate to be so redundant here, but my twenty-six years' professional experience in communications and consumer electronics leads me to conclude that most electronics buffs don't have the patience to bother getting a Service Manual. Instead, they take a rash course of action by jumping headfirst into the innards of their equipment and then crying out for help that they might otherwise not needed. As a "radio doctor," I've spent a lot of time helping others resolve their errors of impatience, but I really would like to spend more time turning my wheels in a forward direction like developing additional modifications, also exploring all of the new equipment that's arriving on the market. I enjoy offering help when its needed, but get the Service Manual-- please!

Next, you need at least a simplistic understanding of electronic terminology and parts. It doesn't have to be much, but you should at least know the difference between a resistor and a capacitor. It wouldn't hurt to know a bit about integrated circuits (IC) and transistors-- like what they look like, their basic functions, and how the pins are arranged. If you agree that such information would be an asset, you can find it in any number of readily available books for electrinics beginners. Check any communications shop's book rack, or the Radio Shack catalog.

Mostly, you'll need a great deal of patience, the ability to read and follow directions. A basic set of electronic hand tools (as shown in Table 4-16-3, the Tools and Materials List for the 6,400 Channel Memory Modification) is highly recommended. Of the tools needed, the **most** important is the one used for soldering. If your soldering iron has a Louisville Slugger baseball bat for a tip, you're going to have serious problems.

Soldering Equipment. A soldering iron for the mods in this book will be rated at 25 to 40 watts, and have a reasonably long and slender body with an "iron-clad" (**not** copper) tip. The tip of the soldering iron should be about an inch long and rather thin with a tiny, flattened end. I happen to prefer Ungar or Weller soldering equipment, and if you can find either of the two brands, then I recommend it to you. I don't care for the Radio Shack soldering guns as much as Ungar or Weller units, but the Radio Shack units are acceptable, reasonably priced and are easier to come by than Ungar or Weller equipment in many areas. Table 4-2 shows the Radio Shack soldering equipment that would make a suitable solder station:

Table 4-2

ACCEPTABLE SOLDERING STATION EQUIPMENT

DESCRIPTION	CATALOG #
Cool Grip Handle, 700°F	64-2080
800°F Heating Unit (33 watt)	64-2081
Iron clad, light duty tips	64-2089
Solder iron holder & cleaner	64-2078
Desoldering braid	64-2090
Desoldering tool	64-2120
Rosin core solder, .032"	64-005 or 64-009

Professionals recommend **Low Wattage** soldering irons and low heat. That's great for the pro's, but novices at soldering are better off with plenty of heat. This is because lower wattages require a critical technique, whereas a bit more heat is forgiving of incorrect technique. However, too much heat can cause no end to grief, tears, gnashing of teeth, and four-letter words, so here are some tips on the technique that you'll want to heed:

1. Use an "iron-clad" tip on your soldering iron.

2. Keep a **sopping wet** sponge nearby to periodically wipe the tip clean. In fact, wipe the tip and then lightly tin the tip with fresh solder immediately before making each solder connection. Do not fail in or change this procedure!

3. A complete soldering cycle goes something like this:
    A. Wipe the tip on a wet sponge.
    B. Apply a bit of fresh solder to the tip, just enough to melt and cover tip.
    C. Touch the tip of the soldering iron to the circuit connection to be soldered. A fraction of a second later, apply solder either to the joint only, or simultaneously to the joint and the tip of the iron. Never apply solder only to the tip when heating a connection.

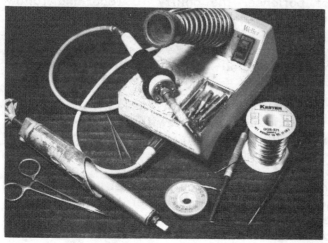

An efficient soldering station and accessories is an important aid to anybody embarking on a career of modifying their scanners, or working on any electronics gear. Good soldering technique also helps. See text.

D. Remove the tip from the connection a fraction of a second **after** solder has flowed into the connection.

Note: If you are doing things right, the maximum time the soldering tip should be in contact with the connection is about three seconds. One to two seconds is possible after you get good at it. In addition to minimum time, the important thing is that the solder must flow or soak into the connection like water into a sponge. Something is very wrong if it takes too long or if the solder doesn't flow right. If you continue doing something wrong, there will be a price to pay as the heat from the soldering iron can damage or destroy equipment components.

If you are unable to do the modifications, or just don't wish to do them, or if you'd like to do them but can't easily round up the necessary parts and materials, at the end of this chapter there is a section titled, "If You Can't Do The Work Yourself..."

OK, you have a Service Manual, an adequate soldering station, basic hand tools, and the will to go forth. Unless otherwise specified, for safety's sake, remove the scanner from from its power source before removing the chassis from the cabinet, and while performing any modification work inside the unit. Even when turned off, a serious shock hazard can exist at certain points in a scanner or receiver's circuitry so long as it remains connected to a live power source. Remove it from that power source (and disconnect the antenna, too) while you are doing any modifications, unless you are specifically instructed to do otherwise.

MOD-1 Restoring the "Lost" Cellular Telephone Bands in The Realistic PRO-2004 and PRO-2005

Preliminary Discussion. The functional circuitry of the PRO-2004/2005 was probably committed to design back as early as 1985. At that time, there was little or no focus on Cellular Mobile Telephone (CMT) from the world of scanners. JIL, Regency, ICOM, and Yaesu had receiving equipment capable of picking up the 800 to 950 MHz Band Mobile bands, and Radio Shack probably felt that it should also cover 800 MHz to be competitive. One thing is for certain, the Realistic PRO-2004/2005 were designed to receive the CMT bands of 825-845 MHz and 870-890 MHz. In fact, the 1987 Radio Shack catalog, prepared in late summer of 1986, specifically proclaimed that the PRO-2004 offered "Continuous tuning from 25-520 MHz and 760-1300 MHz."

By late 1985, the CMT industry had launched a campaign to make it illegal for hobbyist radios to be capable of receiving these frequencies. Apparently, word was starting to circulate amongst CMT users that their conversations might not be quite as private as they had assumed or perhaps led to believe by the person who had sold them their equipment. You might remember the furor raised by stockbrokers, attorneys, and others who thought that "wireline privacy" was standard for their CMT carphones. In the early days of CMT, it was relatively private because scanners didn't tune up past 512 MHz, the upper edge of the UHF-T band. So, from 1982-85 there really was little scant reason to think that anybody else could possibly be listening in on CMT calls. In fact, those who sold CMT carphones usually mentioned this in the sales pitch, pointing out that this was unlike the old 152 MHz and 454 MHz carphones that had been heavily monitored for years by sanner owners.

But then Yaesu, ICOM, and Regency changed all that. By 1985, there were several models of hobbyist hardware on the market that could easily tune the CMT frequencies. Word of this circulated through the CMT and monitoring with expected degrees of horror and delight. That's when the CMT industry shifted into high gear to get the Electronic Communications Privacy Act (ECPA) made into a law to forbid such monitoring. Although scanner users and some from within the manufacturing camp protested the proposed law, it really had relatively little opposition. Even the Electronic Industries Association (EIA) jumped on the bandwagon in complete support. The EIA's strong support surprised many since it is an industry-wide organization comprised of manufacturers in many areas, including those who make scanners as well

In the PRO-2004, diode D-513 was used to lock out the cellular bands. Refer to the diode's location as shown, either above the board or below. When you find D-513, just clip one of its legs. The wires seen in this photo were added for another modification (MOD-16) and won't be seen in an unmodified scanner.

as CMT's. Perhaps the reason was that the revenue-producing potentials of the CMT market outstripped those of the scanner market by about 1,000-to-1. Just a guess. Anyway, the ECPA was rubberstamped through Congress and signed into law in October of 1986.

Thanks to the whatever howl of outrage did come out of the communications hobby (including the ARRL), and some realistic advice on the prospects of enforcing such a law, the ECPA was modified and toned down a bit before it became law. As a result, it is still perfectly legal to manufacture, distribute, own communications equipment that can tune anywhere in the 800 to 900 MHz portion of the spectrum, including those frequencies used by CMT's. Mostly, the ECPA doesn't permit you to intercept (listen to) most conversations (voice or otherwise) which, at any point, is carried over a wireline. CMT's interconnect with the wireline telephone system; ergo, it is a violation of the ECPA to monitor CMT calls. However, the ECPA is a most curious document that was obviously created by politicians having a curious, incomplete, and often cockeyed perception of the very many factors which would have had to have been taken into account to produce a law that didn't contain contradictions, meaningless points, loopholes, and points that are so vague as to make the whole thing preposterous. Still, the gist of it is that you aren't allowed to monitor CMT calls, and it is the law until and if it ever gets wiped off the books. Your best bet is to obtain a copy of the entire law, read it, and obey.

Keep in mind that the PRO-2004 was rolling off the assembly lines in 1986 at the same time all of this crazyness and hullabaloo about the ECPA and CMT monitoring was raging. One might speculate that at some point, the people who make decisions at Radio Shack wondered how and if they should act or react to any of this in view of the fact that their product line also included several CMT models. Perhaps they felt that by selling CMT's and also PRO-2004's that could pick up those frequencies, someone from another CMT manufacturer could question if they were trying to dance at two different weddings-- generally considered tacky and gauche. This, or something else like it, definitely gave Radio Shack second thoughts about the CMT bands in the PRO-2004 scanners.

But how to eliminate this coverage? Units had already been manufactured and possibly were even on the shelves at the warehouse in Fort Worth. Some of this is conjecture, but we know one thing for certain; the arrival of the PRO-2004 at at local stores was delayed for several months, and when they did show up the CMT bands had been blocked out! Still, I have heard from a number of early purchasers who got PRO-2004's that escaped such castration. That didn't happen in my area, though. I went to several local Radio Shacks almost daily looking for early "pre-release" models that might have accidentally gotten through before the big recall and retrofit. No luck, however, no stores had one in stock until all of the modified units were released.

So here we go now, the PRO-2004 and then the PRO-2005, which are essentially the same design, possess the same inherent capabilities to receive all 800 to 900 MHz band frequencies, and mostly look a bit different from one another. All you have to do to restore your PRO-2004 or PRO-2005 to its original designed capabilities is clip a leg of one diode or remove it altogether. Here's how it's done:

PRO-2004. The exact procedure depends on whether you have an "early" unit or a later one. Those units produced before the big recall have the diode retrofitted in place, probably in the USA, however those produced after the recall have the diode installed by the factory in Japan. It's easy to tell (assuming that it hasn't had this CMT restoration done by a previous owner).

1. Disconnect power from scanner and remove the outer metal case.

2. Turn the radio upside down and locate the metal, box-like CPU compartment, marked "PC-3" on the circuit board. Pry off the metal cover of this compartment. If there is any doubt, PC-3 is the sub-chassis that has the **Restart** switch behind a little hole in the rear panel.

3. Refer to the photographs and examine the component area within the CPU compartment. You'll see a large 64-pin IC chip (the CPU). Look to the left of it about two inches and you'll see a row of diodes and resistors parallel with the CPU chip. Look to the top of that row of diodes and find "D-512" and "D-515." Between these two diodes are two pairs of solder pads. Though these are not marked on the circuit board, they are for "D-513" and "D-514," in order.

> A. If a diode is not present in the spot for D-513, go directly to Step 3-B. If D-513 is present, your unit is a later model, and your work is about done. Clip one leg of D-513, and spread the cut ends apart so they can't touch. Replace the compartment cover and the case of the radio. Presto! Your PRO-2004 now has restored the two 800 MHz bands for which it was designed, but which the manufacturer changed his mind about.
>
> B. If D-513 is not visible from the top, then your unit is an "early" model and you have only a little more work to do.
>
> > 1. **Carefully** loosen and remove the 9-pin cable connector, CN-501, from the sub-chassis. CN-501 is located toward the left-rear side of PC-3.
> >
> > 2. Remove the seven screws that hold PC-3 down to the chassis. Lift up and tilt the PC-3 subchassis so you can see the bottom of the circuit board.
> >
> > 3. Somewhere on the center-bottom side of this board will be a solitary diode. Don't worry, because there won't be any other components on this side of the board. Clip one leg of that diode or remove it altogether. You're about done now.
>
> C. Reinstall the PC-3 board and tighten the seven screws. Replace the CN-501 connector. Replace the metal cover of the CPU compartment. Replace the metal case of the scanner. Bingo! Your PRO-2004 now has the frequency receiving capabilities for which it was designed.

Keep in mind that it is not illegal to own receiving equipment covering these or any frequencies, however it is a violation of the ECPA to monitor CMT conversations.

PRO-2005. PRO-2005 is electrically identical to the PRO-2004, although it is physically different. The procedure is the same as for the PRO-2004 except that the components are labeled differently. There are no "early" or "late" units.

1. Disconnect the scanner from AC or DC power and remove the two screws on the rear that secure the top plastic outer case. Remove the top cover by sliding to the rear and lifting up.

2. Locate the Logic & Display board which is immediately behind the front panel display and keyboard. Examine the Logic & Display board in the area directly behind the numeral "3" and "6" keys and you'll see spots for four diodes with two actually installed. The upper diode is marked "D-503" and the lower is marked "D-502."

3. Clip one leg of D-502 or remove D-502 altogether. <u>Voila!</u> Your PRO-2005 has the frequency capabilities for which it was designed.

4. Replace the top cover and screws.

Channel Spacing in The CMT Bands. It's interesting to note that, unlike the channel spacing arrangements in any other radio service, the allocated channel spacing between CMT frequencies is 30 kHz. This is of special relevance since the **Step** function on the front keyboard controls of the PRO-2004 and PRO-2005 lets you set the units for searching in 5 kHz, 12.5 kHz, and 50 kHz increments. Nevertheless, the CPU was factory-programmed to step in 30 kHz increments when searching the CMT bands, and early promotional materials issued for the PRO-2004 mentioned this. For example: Program the lower and upper "Search" limits as follows-- 880.650 and 889.980 and then begin the "Search" mode. You'll immediately notice the "30 kHz" marker show up on the LCD display.

Now, press the front-panel "Step" key and you'll see "5 kHz" displayed. Press "Step" again to see "12.5 kHz" and then again to see "50 kHz." Now press "Step" one more time, and instead of the "30 kHz" you might have expected to see, the display reverts back to "5 kHz." To get the 30 kHz step rate back, you have to press the front-panel "Reset" key. The point here is that 30 kHz stepping is controlled by the CPU, not the keyboard, and is available only for the CMT bands. Any time you might become confused about this when searching, just press the front-panel "Reset" key.

Note: Don't confuse the front-panel keyboard "Reset" key with the rear panel "Restart" key. "Restart" clears memory, but "Reset" simply returns control of the "Search" mode step-rate back to the CPU. The names are similar, don't mix them up.

MOD-2 **Speeding Up The San & Search Rates in The PRO-2004 & PRO-2005**

Method #1

The factory specification for "Scan" and "Search" speed in the PRO-2004/2005 is 8 channels or steps per second in the slow mode, and 16 channels or steps per second in fast. This is not particularly fast; in fact, it is rather slow for many applications. Fortunately, there is a 25% faster speed built in to these scanners which, for some reason, Radio Shack decided not to use-- perhaps saving it for a future model refinement. Actually, there are two different ways to speed up these scanners, and you can do one, the other, or both. See MOD-3 for the second method. This modification is quite simple.

PRO-2004 Speedup #1. The power should be disconnected from the scanner. Located in the same diode/resistor matrix as D-513 described in MOD-1 for the frequency restoration modification, you will find an empty spot for "D-514." It is not marked on the circuit board, but is the first pair of empty solder pads just to the rear of D-515,

The speed-increase diode in the PRO-2004 is called D-514. This is an unmarked spot between D-513 and D-515. If you install a 1N914/1N4148 diode here, you get a free 25% speed increase in scanning.

which is marked. Make sure you have this pair of solder pads identified, then proceed as follows:

1. Install a 1N914 or 1N4148 silicon switching diode in the unmarked spots for D-514. These are in the Radio Shack catalog as #276-1122 and #276-1620. Facing the front of the scanner, looking down into the CPU compartment, the anode (+) leg of the diode will solder to the left pad of the pair. The cathode (-) leg of the diode will solder to the right leg of the pair. The banded end of the diode is the cathode.

2. Soldering the diode in the spot for D-514 can be accomplished without removing the CPU board (PC-3) from the chassis. The solder spots for D-514 are already soldered, so if you position a leg of the new diode on one of the solder pads and then heat the leg of the diode, it will melt the solder of the pad and slip right in. <u>Don't let it slip in more than about 1/16"</u>, though. Repeat this procedure for the remaining diode leg and solder pad. <u>Ensure correct polarity</u> (+) and (-) as described in Step 1.

Comments: That's all there is to this. Your PRO-2004 will now operate in "Scan" and "Search" modes at 10/20 channels or steps per second. There is no drawback to this speedup-- other operations will not be affected.

PRO-2005 Speedup #1. The procedure for the PRO-2005 is electrically identical to the PRO-2004, differing only in the physics and mechanics of the operation.

1. The power should be disconnected from the scanner. Locate the Logic & Display board as discussed in MOD-1, and you will find an empty spot for "D-501." It is not marked on the chassis, but D-503, D-504, and D-502 are marked, so the correct spot is the only one left. Note: D-502 was snipped to restore CMT frequencies, and D-504 should be empty: it will delete 66 to 88 MHz coverage if installed.

2. The correct spot for D-501 is the pair of unused solder pads between D-503 and D-502. Observing correct polarity, install a 1N914 or 1N4148 silicon switching diode (Radio Shack #276-1122 or #276-1620) in the unmarked spots for D-501. The anode (+) leg of the diode will solder to the inner unused solder pad, while the cathode (-) leg of the diode will solder to the outer unused solder pad.

Comments: This is the complete mod. The PRO-2005 will now "Search" and "Scan" at a rate of 10/20 channels or steps per second. Other than the increased speed, other functions will remain unchanged.

MOD-3 Speeding Up The Scan & Search Rates In
The Pro-2004 and PRO-2005

Method #2

PRO-2004 Speedup #2. To my way of thinking, even 10/20 channels per second is pretty slow. Fortunately, there is another relatively simple way of speeding things up by an additional 33% (15/30 channels per second). You will need to acquire a quartz crystal-- not the kind used for curing your spiritual imbalances, but one cut for a frequency anywhere between 9.0 and 10.0 MHz. You can try a crystal cut higher than 10 MHz, but it might not work. Commonly available microprocessor crystals are well suited for the purpose. You can even order one from Radio Shack. Again, this modification is simple and easy, but there's <u>one</u> minor drawback.

You could find that 10.0 MHz is <u>too fast</u> for your scanner, or the crystal you select might not be of the correct type, and if so, things won't work. That happened to me once, although I have never heard of this limitation from the many people who tell me that 10 MHz crystals worked fine for them. In the case of my PRO-2004, it was no problem because I operate a communications service shop and happen to stock hundreds of crystals. So, when 10.0 MHz didn't work, I tried others until I found that 9 MHz worked just fine. I have subsequently tried several 10 MHz microprocessor crystals and they all performed, so success probably depends on the type of crystal, too.

You may run into this problem and have to try several crystals before finding one that works. Still, there is no problem in reversing this procedure, so give a try for **warp speed** as follows:

1. The power should be disconnected from the scanner. Locate the CPU clock "ceramic oscillator," CX-501. It's easy to find. Pry off the metal cover of the CPU compartment, PC-3, and look in the right-rear corner of that compartment. Just to the right of IC-502 will be what you're seeking, CX-501-- a little plastic gizmo sticking up out of the circuit board. It will be a blue three-legged device, probably marked "CST7.37MT" or similar.

2. You're better off raising the CPU up from the main chassis for this mod, so loosen and disconnect the 9-pin cable connector (CN-501) from the subchassis. CN-501 is located toward the left-rear side of PC-3.

3. Remove the seven screws that hold PC-3 down to the chassis. Lift up and tilt the PC-3 subchassis so you can see the bottom of the circuit board.

4. Identify the three solder spots that hold CX-501. Desolder those three spots with desoldering braid or a vacuum solder sucker. Remove CX-501 and store it away in a safe place.

Location of the 10 MHz Speed Increase Crystal in the PRO-2004. The view shown is the rear-right corner of PC-3.

5. Ignore the center solder spot from where the CX-501 had been removed. In the two outer solder spots where CX-501 was removed, install a quartz crystal cut for a frequency between 9 and 10 MHz.

Warning - Caution - Important: If your crystal has a bare metal shell, wrap a layer of tape around it. You might be inclined, as I did, to lay the crystal flat-- and if you do, it will rest on top of a bare lead of D-51 just below. The bare metal of the crystal case will cause a short circuit at that point. The short circuit will cause IC-9, a CMOS voltage regulator, to instantly shift into self-destruct mode. Replacing it is definitely less fun than than a barrel of monkeys. But, I plead gross ignorance and sheer stupidity. Now that you've been warned, you don't get to use the same excuse.

6. Reinstall the CPU board and tighten the seven screws. Replace the metal cover and test the radio.

Test Procedures. First determine that the radio works by selecting a known station, such as the closest NOAA 162 MHz band weather transmitter, or your local police department's main frequency. If you don't hear anything, then maybe the crystal you used is a bit too high in frequency or it's the wrong type. Try a different crystal, or one cut for a slightly lower frequency.

Once you confirm that the scanner will receive, disconnect the antenna from the radio. Flip the "ATT" switch on the rear of the scanner to the "10 dB" position. Turn the "Squelch" control all the way clockwise (CW). Press "Manual : 1 : Manual". Holding a stopwatch, hit the "Scan" button and start the stopwatch simultaneously. Stop the watch at the instant the scanner passes Channel 300 (or 400, if that's how many channels you have. See MOD-15.)

Perform the following calculation:

$$\text{Speed} = \frac{\text{\# of channels scanned}}{\text{Elapsed time in seconds}}$$

Mine worked out this way: Speed= 400 channels divided by 14.8 seconds = 27.03 channels per second. Slow speed is half the speed of fast.

When you have completed the test procedure, return the "ATT" switch to its original position prior to the test, or else the scanner will be operating with reduced sensitivity.

By the way, my 2004 was one of the first to hit the market. If you are successful with a 10 MHz crystal, your speed will probably be closer to 30 ch/sec than I got.

PRO-2005 Speedup #2. The speedup for the 2005 is electrically identical to that of the 2004. The differences are in physics and mechanical layout only. Acquire a quartz crystal with a frequency between 9 and 10 MHz. Install the crystal in place of the PRO-2005's ceramic oscillator, CX-501, which is identical to that discussed above for the PRO-2004.

In the PRO-2005, CX-501 is located in the center area of the Logic & Display board. When the modification is completed, use the same test procedure as given above for the PRO-2004 Speedup #2.

MOD-4   **Improving The Squelch Action In**
                                     The PRO-2004 and PRO-2005

This modification was highly touted when it was first introduced several years ago. Every magazine and little newsletter raved about it. I even wrote one of the articles on how to do the mods. These days, I'm not quite as excited about the mod, but it is worthy of mention. By the way, the problem about to be discussed and solved is common to all current Realistic brand scanners, and the "fix" is identical to just about all.

First, find a vacant channel where there are no signals to interrupt things. Now,

Here's where to look for R-148 in the PRO-2004. This is where you'll want to work to modify the operation of the squelch in the PRO-2004. In the long run, it might be best to not fool with the squelch, despite its apparent shortcomings.

with the "Squelch" turned all the way counterclockwise (CCW), slowly rotate it clockwise (CW) and stop at the precise point where the static sound in the loudspeaker is silenced. That point is rather abrupt, and you might want to do it several times just to familiarize yourself with when this occurs.

Now, even more slowly, turn the "Squelch" counterclockwise (CCW) and you'll immediately note that the sound does not return until the knob has been turned backwards by a small, but significant, amount. That's the problem! "Purists" feel that certain feeble signals will go undetected because of this sloppy or rubbery squelch action. The hue and cry is for the "Squelch" to come "on" and "off" as close to the same point as possible. That's fine, theoretically, because that's the way it should be.

From a practical standpoint, however, with a "Squelch" made that critical, the merest noise pulses and bursts would open the "Squelch" so often that you'll go crazy hearing "machine-gun" bursts of static all the time. The scanner will stop as much on static and noise as it will for signals, and the question is if this is a good trade off for getting razor sharp squelch action. I find it annoying and disruptive to serious scanning.

Here is what the deal is, at least apparently. Radio Shack realizes that its customers are mostly kind of like me in the sense that scanning for noise and static is about as enjoyable as watching the "snow" on a vacant TV channel. So, a small amount of "overlap" (more correctly called "hysteresis") was deliberately designed in to eliminate an erratic "Squelch." Even so, some "purists" feel their caution is a bit overdone. Below is a modification to correct it, and my own innovation provides an internal adjustment for you to set as you like it best. If you don't care and want a squelch so critical that you hear every snap, crackle, and pop, then set it that way for yourself. If you want just a little rubber and slopover, fine...adjust for that. OK?

PRO-2004 Squelch Improvement. Disconnect the power from the scanner. Remove the outer case of the scanner. Facing the front of the radio with it positioned upright, look down into the center area of the receiver board.

1. Locate the subchassis with a metal cover that is partly hidden under the sloping faceplate of the radio. There are thirteen holes in this metal cover for the purpose of adjustments beneath. Carefully pry off the cover of this subchassis. That printed circuit board is marked "PC-1" but you can't see it because it is hidden under the sloping front panel.

2. Locate IC-2, which will be in the far-left side of the PC-1 subchassis, and off by itself, away from other IC's. The type number of IC-2 is TK-10420, though the "TK" may not be marked on the chip in your scanner. IC-2 is located just to the left of a crystal, X-2.

3. Locate R-148, a 47K (color code: yellow/violet/orange) resistor between Pins 12 and 14 on the left side of IC-2. It is clearly marked on the circuit board.

4. Tin with solder the top leg of R-148 so that the paint is scorched off and the metal wire is exposed. Cut the top leg of R-148 somewhat away from the body of the resistor so that both cut ends can be soldered later. Spread the two cut ends apart.

5. Between the two cut ends of R-148, solder a 200K to 500K variable trimmer potentiometer.

You're done, except for adjusting the trimmer pot and replacing the metal cover and case. Following is the theory of the modification (in case you're curious) plus a note or two to help you along.

Theory: Resistor R-148 provides 47,000 (47K) ohms between Pins 12 and 14 of IC-2. This "programs" IC-2 to provide a particular squelch action. Less resistance than 47K, and the slop and rubbery action gets worse. The higher the resistance, the tighter the squelch action. Some purists simply remove R-148 so that there is infinite resistance between Pins 12 and 14. Personally, I find a total resistance of 150K to 200K to be better. Therefore, since 47K is already in place, you need an additional 100K-150K in series with R-148. The trimmer pot allows you the freedom to decide what you like best, and change your mind from time to time. If you don't have a trim pot, you can always solder a 100K resistor between the two cut ends of R-148 and still get some improvement over the stock squelch action.

PRO-2005 Squelch Improvement. For the PRO-2005, you'll first disconnect the power from the scanner then locate IC-2 (TK-10420) and resistor R-152, which happens to be 33K ohms (color code: orange/orange/orange). R-152 runs between Pins 12 and 14 as usual, but is a surface-mount type and is installed on the bottom side of the board below where IC-2 is installed. You'll have to remove R-152 from the underside of the board and then you can work from the top after that. Be careful when soldering to Pins 12 and 14 of IC-2. It can be done, but be careful. A total of about 150K ohms between Pins 12 and 14 should be just about right.

Other Realistic Scanners. All Realistic brand scanners since the old PRO-2002 use the same TK-10420 or generic equivalent which provides a lot of the receiver functions on a single chip. Pins 12 and 14 program the squelch action, so by increasing the resistance between those two pins, you'll cause the squelch action to become tighter. This is true of the PRO-2002, PRO-2021, and all other Realistic scanners of recent times. Find the TK-10420 chip; find Pins 12 and 14; locate the resistor between those two pins; increase that resistance by 100K to 200K, and you're in business.

MOD-5 Improved Tape Recorder Output Quality In
The PRO-2004 and PRO-2005, Plus Other Scanners

Ever notice the harsh and disagreeable quality of tape recordings made from the "Tape Out" jack on the rear of the PRO-2004/2005? I don't even like the quality of the sound that comes from the recorder outputs of most scanners, communications receivers or even CB transceivers. The audio signal at the output jack is at a very high level and, to make matters even worse, it's noisy and distorted. Part of the problem is that modern tape recorders work just fine with "line level" signals, and that means levels of only a few millivolts. Therefore, the high level signals at the "Tape Out" jack come out like the Marines arriving at the Halls of Montezuma-- they overdrive most recorders and cause distortion. A quick review of the schematic diagram led me to spot the problem and then see a handy solution.

For my PRO-2004, I left intact the stock "Tape Out" jack and its circuit, but I also added a separate "Tape Rec Out" jack and circuit for "line level" quality output. The circuit requires three parts, and a bit of shielded microphone cable. Refer to Figure 4-5-1 for the wiring diagram, Table 4-5-1 for the Parts List, the photos and step-by-step instructions that follow here.

Figure 4-5-1
Improved Tape Recorder Output

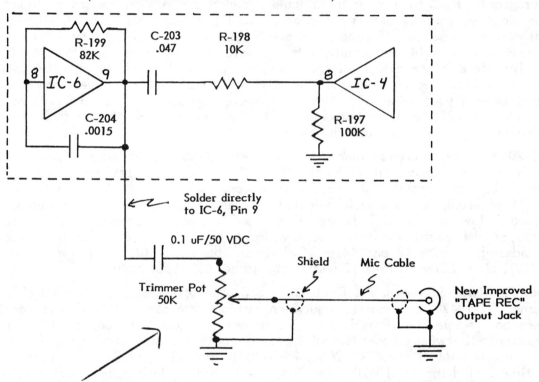

Build this simple circuit to improve the "TAPE REC" quality from any scanner.

<u>Notes - See Text</u>

1. Adjust trim pot for best quality tape recordings.
2. PRO-2005: Solder the .1 uF capacitor to IC-5's Pin 13.
3. Install circuit inside scanner. Use shielded microphone cable from trim pot to the new "TAPE REC" output jack.
4. For other scanners, solder the .1 uF capacitor to the "Detector" output.

Improving the tape recording quality requires the installation of some parts.

Table 4-5-1

PARTS LIST FOR IMPROVED TAPE RECORDER QUALITY

QUAN	DESCRIPTION	RADIO SHACK #
1 ea	Trimmer potentiometer; 50k ohms	271-219 or 271-283
12" approx	Shielded Microphone cable:	278-752
1 ea	.1-uF/50 VDC capacitor:	272-135 or 272-1069
1 ea	RCA Phono Jack:	274-346
	or 1/8" phone jack:	274-251
	or your own choice of jack	

PRO-2004. Disconnect scanner from its power source, then:

1. Locate IC-6 on the main receiver board. You shouldn't have to pry off any metal covers other than to remove the main case at the beginning. IC-6 is located at the left-rear quadrant. IC-6 is the middle of three integrated circuits situated in parallel in that vicinity of the main receiver board.

2. Locate Pin 9 of IC-6.

3. Clip one leg of the .1 uF/50VDC capacitor from the Parts List so that it is about 1/4" long. Solder that leg of the capacitor directly to Pin 9 of IC-6. Tin Pin 9 of IC-6 with solder first and then quickly solder the capacitor to that pin.

4. Solder the mini 50K trimmer potentiometer as follows:

   A. One <u>end leg</u> of the trimmer (a stator) to a nearby ground. The nearby metal shield case is a good ground spot.

   B. Solder the free leg of the capacitor to the <u>other end leg</u> (the other stator) of the trimmer pot. Put a bit of insulation sleeving over this leg of the capacitor before soldering it to the trimmer pot if it has a chance of touching anything.

5. Prepare a short length of shielded microphone cable and connect one end as follows:

   A. Center conductor to the center leg (the rotor) of the trimmer.

   B. Shield conductor to the nearby ground spot where one end of the trimmer pot was soldered.

6. Connect the other end of the microphone cable to a jack of your choice mounted anywhere you like on the rear chassis or the front panel. The shield at this end of the mic cable does not absolutely need to be connected to ground, though it is advisable to do so for ultimate quality.

Note: My new "Tape Rec Out" jack is on the front panel near the headphone jack. I chose that location so it could be used in tandem with the automatic tape recorder switch described in MOD-6 of this book. If you are not sure where to locate your new "Tape Rec Out" jack, read MOD-6 first for some ideas.

7. Connect a tape recorder to your "Tape Rec Out" jack and tune in a station on the scanner. Adjust the trimmer potentiometer for best quality recording with a minimum of noise. Note: Turning the trimmer pot all the way to one end will result in either too strong signals or too much noise, or both. At the extreme other end of the trimmer, there will be zero signal and zero noise. Somewhere between these two extremes is one ideal spot you can find by experimenting.

PRO-2005 Note. Follow these same steps of procedure given for the PRO-2004, above except that you'll work with IC-5, Pin 13 instead of IC-5, Pin 9. IC-5 in the PRO-2005 is located on the main receiver board pretty much as described above for IC-6 of the PRO-2004.

MOD-6 **Automatic Tape Recorder Switch for The PRO-2004, PRO-2005 & Other Scanners, Too**

Voice Activated (VOX) tape recorders have come a long way in a short time, and they do work to some extent for scanner enthusiasts, but there can be more trouble than the VOX feature might be worth. Depending upon the scanner, sometimes even when the "Squelch" is set and no signals are coming in, there is a substantial amount of noise in the "tape out" or "headphone" circuits. This is notable in both the PRO-2004 and PRO-2005. This noise triggers the VOX circuits of the tape recorder, and the result is as if there were no VOX at all. The tape recorder grinds away, recording the sounds of silence plus a little noise. For those situations that do work, the precise VOX settings are critical and not highly reliable.

To avoid all such problems, here is an automatic tape recorder switch that is easily and enjoyably built. It controls a tape recorder so that when the "Squelch" is set, the recorder will auto-stop and just sit there waiting until a signal breaks the squelch. Then the tape recorder will auto-start and record the signal until it leaves the air. In that manner, a 45-minute or a 120-minute tape can be stretched to the max with no dead spots between the action.

This one works just fine, and the cost is minimal. Operation is triggered by the "Squelch," not by audio, so it is relatively fail-safe and foolproof. If the signal is strong enough to break the squelch, it will get captured on your tape. Built on a "perf board" with a handful of components, the Automatic Tape Recorder Switch will assure you of not missing the action on your favorite channel (or group of channels) while you aren't present. Check the Parts List (Table 4-6-1), see the wiring diagram (Figure 4-6-1), then follow the directions.

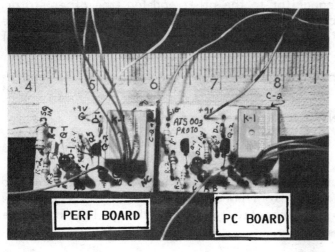

Automated tape recorder switch layout. The component side of the perf board (photo left) was first covered with a layer of plastic before the parts were stuffed into the board to facilitate marking of the labels.

Figure 4-6-1
Automatic Tape Recorder Switch

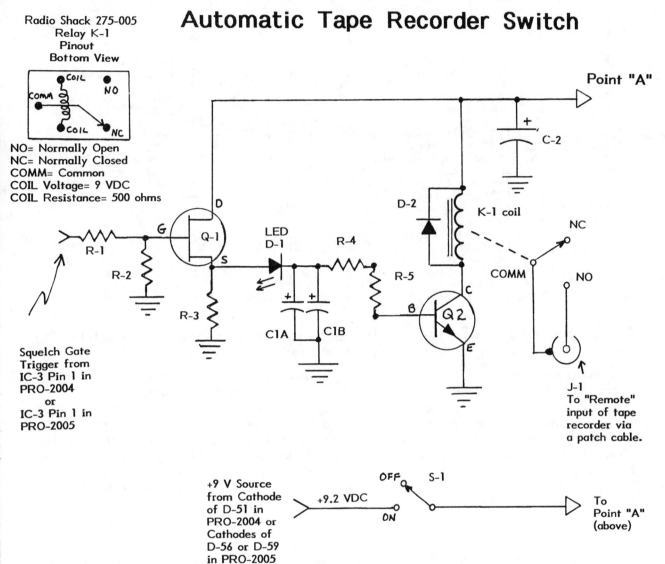

Notes - (See Text)
1. C1A & C1B should be 32-33 uF (total) tantalum capacitor.
2. D-1= Any LED with forward voltage drop off of about 1.6-1.8 V will do.
3. D-2 can be substituted with a 1N914/1N4148 switching diode.
4. R-1, R-3, R-4, & R-5 can be varied.
5. Q-2 can be any NPN switching transistor (ECG-123A, etc.).
6. Maximum current drain is about 20 ma.

Table 4-6-1

PARTS LIST FOR THE AUTOMATIC TAPE RECORDER SWITCH

Ckt Symbol	Quan	Description	Radio Shack #
C-1a	1	22-uF/16 VDC tantalum capacitor:	272-1437 See NOTE[1]
C-1b	1	10-uF/16 VDC tantalum capacitor:	272-1436 See NOTE[1]
C-2	1	2.2-uF/35 VDC tantalum capacitor:	272-1435
D-1	1	LED;	276-026
D-2	1	1N4004 diode;	276-1103
J-1	1	Jack of your choice; suggested:	274-292
K-1	1	SPDT mini relay; low current drain;	275-005
Q-1	1	2N3819 FET transistor;	276-2035
Q-2	1	MPS-2222A transistor;	276-2009
R-1	1	220k, 1/4-watt resistor;	271-1350
R-2	1	10-meg, 1/4-watt resistor;	271-1365
R-3	1	4.7k, 1/4-watt resistor;	271-1330
R-4	1	10 k, 1/4-watt resistor;	271-1335
R-5	1	3.3 k, 1/4-watt resistor;	271-1328
S-1	1	SPST mini toggle switch or equiv;	275-624 or 275-612
Misc		Perf board, approx 1.25" x 1.25";	276-1395

NOTE[1]: C-1a and C-1b tantalum capacitors are paralleled to act as one 32-uF capacitor, C-1. A single 33-uF/16vdc tantalum capacitor would be perfectly acceptable, but Radio Shack doesn't have it.

PRO-2005 Note. This modification can be easily accomplished using the following procedure, **but first** see the exceptions and differences at the end.

1. Fabricate the Automatic Tape Recorder Switch according to the wiring diagram in Figure 4-6-1. Parts layout is not critical. Be sure that the tantalum capacitors (C-1a, C-1b, and C-2) are wired in correct polarity (+ and -). Also be sure that correct polarity of the two diodes, D-1 and D-2, is observed relative to the wiring diagram in Figure 4-6-1. Check the package the the LED (D-1) came in for information on correct polarity before wiring it in, or turn the LED upside down so that the pins point toward you and you'll see that next to one of the leads at the base there will be a "flat" side of the otherwise circular base. The lead next to the flat spot is <u>usually</u> the cathode (-). Solder color-coded wires about 12" long to: Board ground; Point "A"; the free end of R-1; and to the "Comm" and "N.O." terminals of Relay K-1.

Prelimary testing: You should test the board "out of circuit" once it has been completed. Testing is easy, as follows: Connect the negative (-) side of a 9-volt battery to the ground of the new board. Connect the positive (+) side of the 9-volt battery to Point "A" of the new board. Then, connect the (-) side of a 1.5-volt

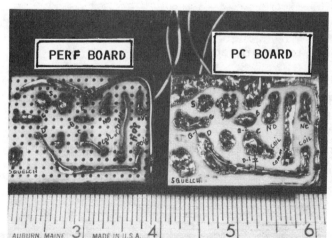

This is the top-to-bottom flip view of the automated tape recorder switch.

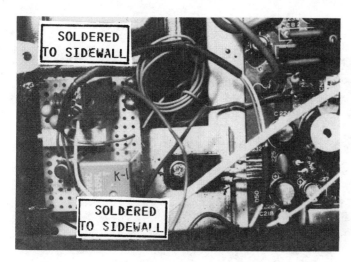

Automatic tape recorder switch installed in PRO-2004. The circuit board is mounted by soldering two ground lugs to the chassis side wall. Disregard the unfamilar components, this was an early prototype design.

flashlight battery to the board ground, and then the (+) side of the 1.5-volt battery to the free end of R-1 on the new board. At the instant you touch the (+) battery terminal to R-1, you should hear the relay (K-1) click (energize). Then remove the (+) battery connection from R-1 and the relay will click again as it de-energizes.

Note: The circuit might "trigger" without the use of the 1.5-volt battery if you just <u>touch</u> the free end of R-1. Noise is picked up by your body and coupled into Q-1 is sometimes sufficient to trigger the relay.

If everything works OK, move on to Step #2. If not, troubleshooting is required. First try triggering the circuit with two flashlight batteries hooked in series (3 volts), with the negative (-) terminal connected to the board ground, and the positive (+) terminal to the free end of R-1. If the relay does not "click" now, there is definitely a problem on the board. Double check the polarity of the two diodes (D-1 and D-2) against the wiring diagram to make certain it's correct. LED's are very prone to reverse polarity errors. Otherwise, about the only thing that can go wrong are the transistors or, of course, the wiring itself. This is a neat circuit and it works great, so if you've followed directions and your connections work, very little can go wrong.

2. The critical thing is to decide <u>where</u> on the scanner you want the switch (S-1) and the tape recorder control jack (J-1). If you perform the improved tape recorder output quality modification (MOD-5), you may want that "Tape Rec Out" jack to be close to the tape recorder control jack (J-1) of this mod.

Install S-1 and J-1 in locations of your choice. Put a ground solder lug on J-1 (if it doesn't already have one) before installing and tightening the lock nut.

3. The completed circuit board should be mechanically installed somewhere handy inside the radio so that it doesn't just flop around loose. I installed mine by soldering the ground leg of the board directly to the sidewall of the inner metal main chassis near the scanner's power supply. The inter-wiring in Steps 4 through 9 can be done either just before or just after the board is mounted in the place of your choice.

4. Connect the hookup wire from the free end of R-1 (220K) on the new board to IC-3, Pin 1 in the PRO-2004. IC-3 is located in the left-rear area of the PRO-2004's main receiver circuit board. Note: This will be the squelch gate trigger for the new board.

Helpful hint: There is a circuit trace that leads away from IC-3, Pin 1 to an unused solder pad. Solder the hookup wire to this pad instead of directly to Pin 1.

5. Connect the hookup wire from Point "A" on the new board (junction of D-2, C-2, drain of Q-1 and the coil of K-1) to one terminal of switch S-1.

6. Connect a hookup wire from the other terminal of Switch S-1 to the cathode (-)

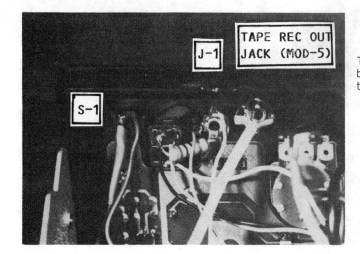

The switch and jacks for Remote and Tape Out can be installed anywhere, but space is available near the headphone jack (as shown).

leg of D-51 in the PRO-2004. Note: The cathode of D-51 is an excellent source of +9.2 VDC in the PRO-2004's power supply to run the new board, and is located in the left-rear quadrant of the main receiver board under the speaker, which will have to be temporarily removed for access below.

7. If you haven't already done so, connect a hookup wire from the ground (common) side of the new board to the metal chassis of the PRO-2004. You could, instead, connect a hookup wire from ground of the new board to any printed circuit board ground of the PRO-2004 as well; either/or. If you solder the new board's ground directly to the metal chassis (as I did), then this step is not necessary.

8. Solder a pair of hookup wires to the "N.O." and the "Comm" contacts of the relay (K-1), if you haven't already done so.

9. Solder the other end of this pair of wires to the hot and ground lugs of the new tape recorder control jack (J-1). See Step 2. The wire from the "Comm" terminal of the relay should go to the ground lug of J-1.

10. <u>Connection/Testing/Adjustment/Operation:</u>

    A. Connect a jumper cable from the tape recorder's "Remote" jack to the tape recorder control jack (J-1) newly installed in the PRO-2004.

    B. Connect another jumper cable from the "Tape Out" jack on the rear of the PRO-2004 (or your new "Tape Rec Out" jack if you've performed MOD-5) to the record input jack of your tape recorder.

    C. Set S-1 to "Off." Set the scanner to an unused channel with the squelch activated (turned up). Set the tape machine to "Record" mode.

    Note: the tape recorder should <u>not</u> actually record yet because S-1 is off and the jumper cable from J-1 to the "Remote" jack on the tape machine pauses the recorder. Now turn S-1 to the "On" position. Nothing still will happen because the "Squelch" has not released.

    D. Now, slowly turn the "Squelch" counterclockwise (CCW) until you hear static from the scanner. At that precise moment, the tape recorder should begin recording.

    E. Now turn the "Squelch" clockwise until the scanner is silenced, and then after a very brief delay, the tape machine should shut down.

If all is well at this point, you're ready to clean up and restore the scanner to its normal operating position. Turn S-1 off when the automatic tape recorder switching device is not in use.

Front panel of the PRO-2004 showing the mounting of the new tape recorder switch (at left) and new TAPE REC out jack from MOD-5.

Tech Theory of The Automatic Tape Recorder Switch. Sometimes it is helpful to know the theory behind why the circuits we build work. Here's what happens in this most useful and neat circuit.

R-1 samples the status of the squelch signal in the scanner. When the "Squelch" is set but no signals are present to break it, IC-2's Pin 13 outputs a 0.0 VDC signal to IC-3, Pin 1. R-1 samples that DC signal at IC-3's Pin 1. When the squelch breaks (like when a signal comes in) or the knob is turned down by the operator, the output of IC-2's Pin 13 rises to +4.5 VDC, again sensed by R-1.

R-1 serves to isolate the new circuit from the PRO-2004's circuitry so that if ever a problem develops on the new board, it won't affect the scanner. R-1 also couples the squelch signal to the gate of Q-1, a Field Effect Transistor (FET) with an extremely high input resistance. It, too will not affect the scanner's circuitry, however when a squelch voltage of some +DC level is applied to Q-1's gate, then Q-1 increases conduction. The voltage drop across R-3 increases, which raises the forward bias of D-1. As soon as the voltage rise appears across R-3, then C-1 charges to that voltage through LED diode D-1. The LED is not used for light, though it may flicker or dimly glow. Charging of C-1 takes place almost instantly,

As C-1 charges, the same charging voltage is "felt" via R4 & R5 to the base of Q-2. That voltage biases the base of Q-2 more positive thereby increasing collector current through the coil of relay K-1. K-1 then energizes or changes state, and makes a short circuit from the "common" terminal to the "normally open" terminal. That short circuit (or, better termed a "contact closure") is passed through output jack J-1, into the jumper cable to the "Remote" jack of the tape recorder. If the tape recorder has been set to the "Record" position (and, if S-1 is on), then as soon as K-1 energizes, the tape recorder will begin to record.

When the squelch activates, a 0.0 VDC signal is returned to the gate of Q-1, thereby decreasing conduction through Q-1. Current through R-3 and the voltage across R-3 decrease. With nothing to charge it now, C-1 must discharge, but it cannot discharge through back through R-3, since diodes conduct only in one direction. Therefore, C-1 discharges through R-4, R-5, and Q-2. The resistance of R-4 and R-5 slows the discharge rate of C-1. As long as C-1 discharges (about 0.5 seconds), Q-2 remains biased for heavy conduction, and relay K-1 will remain energized. Depending on the values of C-1, R-4, and R-5, the discharge time of C-1 will be less than 1 second. When C-1 is discharged, Q-2 decreases conduction. Current through the coil of relay K-1 drops below the holding level required, so K-1 de-energizes and returns to an open circuit to J-1. At the moment K-1 is de-energized, the tape recorder will stop and wait for the next contact closure which can be triggered only when the squelch of the scanner breaks again. The tape recorder will remain in standby until signals break the squelch.

Diode D-2 is a protective component, strategically placed to do nothing unless the coil of the relay generates a high voltage spike (which coils are known to do), in which case, D-2 shorts the spike. This prevents it from causing any damage to the board and the scanner. C-2 assures smooth DC power to the board, filtering out noise and AC hum that might come in from the scanner's power supply at D-51, or which might go back into the scanner from the board.

You can set the scanner to "Manual," "Scan," or "Search" operating mode and than then go to bed, to work, to the store, to the movies, or get busy in some other room of the house. The tape recorder will record only active channels and only when the squelch breaks. There won't be any dead or blank spots on the tape. A 45 to 120 minute tape can deal with hours or even days of activity on a given channel if it isn't too busy. Obviously, a busy channel is going to use up tape more rapidly.

PRO-2005. Follow the same procedure as just shown for the PRO-2004, except:

A. Power the new circuit board at Point "A" from the cathode (-) of either D-56 or D-59, located in the left-rear area of the main receiver board.

B. Sample the squelch signal to R-1 from IC-3's Pin 1. IC-3 is located in the left-rear area of the main receiver board.

Other Scanners. This modification can be readily adapted to most any other scanner that employs a positive squelch "gate." That is, when the squelch breaks, a +DC signal of 2 to 5 volts must be generated, and then fall to near zero volts when the squelch is set. Find that signal point and feed it to R-1 as previously described. Point "A" on the new board must be powered with about +9 VDC, though there is some flexibility, I suppose, of 8 to 10 volts.

Note: This circuit, as designed, energizes relay K-1 when the squelch gate trigger or input voltage rises to about +1.20 VDC or more. The relay will de-energize when the input gate trigger drops below about +0.2 VDC. Negative voltages are of no importance. The main thing is that when the squelch is set, the voltage to R-1 must be less than 0.2 VDC. When the squelch breaks, the voltage to R-1 must rise to at least 1.2 VDC.

Different combinations of R-3, R-4, R-5, and C-1 could compensate for slightly different squelch actions in other scanners. Experiment with R-3 only as a last resort.

MOD-7   Improved Low Visibility Keyboard for The PRO-2004

Sorry, folks, but this one's a PRO-2004 exclusive. The concept given here was originated by Dr. Milan Chepko of Mississippi. Like me, he has a problem seeing and using the PRO-2004's flat, featureless, keyboard in low light levels. Dr. Chepko and I corroborated to work out a minimal-cost, highly effective improvement for the keyboard under low-light conditions.

Here it is, late at night, and you've settled back to do some serious scanning. The lights are dim and your PRO-2004 is chattering away. The 2004 is scanning through several memory banks at warp speed when suddenly something really interesting begins to develop on one frequency. As an experienced operator, you well know that even the "Delay" mode sometimes isn't long enough to keep the set from scanning again before the channel is occupied by another signal. If you don't hop on that "Manual" key in a flash, the scanning action might start up again and you will might never know which channel the action was on. But, with the keyboard almost invisible in the dim light, you probably won't be able to be able to find the "Manual" key in time.

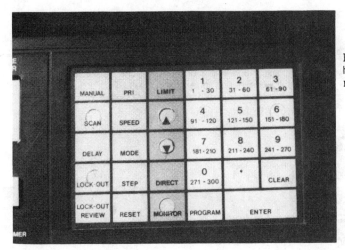

Improved low-visibility keyboard overlay. Note the holes punched in the plastic overlay at certain recognition points.

Same scenario can apply if you're searching unknown bands and you hit upon a live frequency of interest. Before you can gather up your wits (and sights) to hit the "Monitor" key, the whole thing is gone. Familiar?

While Dr. Chepko and I can't increase your reaction time, we can offer a fix to make things easier on your vision. Conjure up a piece of clear Acetate (stiff, but thin, clear plastic) like that used on "document protectors" or photo albums. Cut a piece exactly 95 mm. by 65 mm. Set the clear plastic on the keyboard so that it fits into the recesed rectangle. A bit of trimming might be required for a perfect fit. Better to have it a tight fit than too loose. If loose, it will slide around and annoy you.

Once it fits, use a marking pen to place small dots in the center of certain critical keys, such as "Scan," "Monitor," and the up and down arrows. Now, use a hole punch or sharp razor blade to make small cutouts over each dot you've marked. Don't get carried away and do _all_ the keys; you won't need them _all_ most of the time, and with only a few reference holes you'll better know where everything else is located. For example, with a hole over "Scan," you know that "Manual" is just above that hole, and "Delay" is just below. With a hole on the "Monitor" key, you know that "Direct" is just above, and "Reset" is to the left. The numerals and "Program" can be left alone because you don't program often, and certainly not in low light levels.

So, in dim or no light, and with just a bit of practice, your fingers can find their way around the important areas of the keyboard while your eyes rest. And it still looks great in bright light since the clear plastic is hard to see. The plastic plate you've added will remain in place without glue or other adhesives and yet it may be removed when desired.

By the way, this is an excellent aid for vision impaired scanner fans, but then, they've probably already come up with the idea on their own.

MOD-8   Improved Headphone Audio for The PRO-2004, PRO-2005 & Perhaps Other Scanners, Too

Dr. Chepko also observed that the headphone feature of both the PRO-2004 and PRO-2005 leave some room for improvement. Perhaps this is also true for other scanners as well. Most of the time, the cure is simple. But first, let's identify the problem we are facing.

Plug an earphone into the jack on the front panel of your PRO-2004/2005. Select a signal to monitor and adjust the volume to a comfortable level. Now remove the earphone from the jack. The shock wave of audio decibels that suddenly emerges from your loudspeaker is usually enough to frighten the neighbors. You quickly fumble with the volume control to reduce its setting. You've discovered that it takes a lot more

Improved headphone audio; 10-ohm resistor installed. The right end of the resistor is soldered to the headphone ground lug. The other end of the resistor is soldered to a PC board ground trace below.

volume control for a given level of sound from the headphone jack than it does for either the internal or external speaker. Let's equalize the sound level between the headphone jack and the speaker. It's rather easy to do.

The culprit that's causing the problem is R-218 in the PRO-2004 (R-228 in the PRO-2005). In either radio, there is a 270-ohm resistor connected in series with the ground lug of the headphone jack at one end of the circuit to board ground at the other end. What this does is **nothing** as long as you're listening to speaker audio. When you plug an earphone into the jack, then 270-ohms is in series with the earpiece and this resistance acts as a fixed volume attenuator. That's why you have to crank up the sound to compensate. The fix is easy.

1. Disconnect the scanner from its power source. Remove the case of the scanner and flip the radio upside down so you can examine the headphone jack from inside the front panel. There will be two wires to the jack.

Note: You may wish to remove the front panel from the main chassis to make access to the phone jack a little easier. There are four countersunk phillips screws, two on either side of the front panel that have to be removed.

2. Find the ground lug (the **outer** ring or shell) of the phone jack.

3. Solder one lead of a 10-ohm, 1-watt resistor (Radio Shack #271-151) to the ground lug of the phone jack. Don't cut any wires!

4. Solder the other lead of the 10-ohm resistor to any circuit board ground spot or directly to a metal chassis. You may have to solder a bit of hookup wire to the free end of the resistor in order to enable it to reach a ground point.

That's it. Button things up and enjoy headphone volume that is pretty much equalized with the speaker volume.

What we have done here is to parallel the headphone limiting resistor (R-218 in the PRO-2004, R-228 in the PRO-2005) with 10 ohms to reduce the overall series resistance to the phone jack. Don't monkey with R-218 (or R-228 in the PRO-2005) because it is hidden on the circuit board somewhere and access or removal is neither convenient nor required.

Other Scanners may have the same problem and the fix is identical.

MOD-9 **Disabling the "Beep" in The PRO-2004 and PRO-2005**

They went to all of the trouble to make these sets "beep" when the keyboard keys are pressed, and apparently lots of owners got annoyed! I say "apparently" because I've seen ads in the scanner hobby media from shops from comms shops that offer to

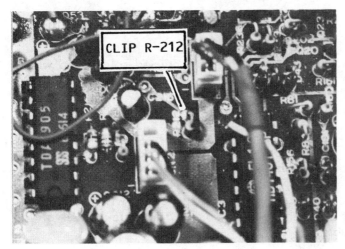

To disable the "beep" in the PRO-2004, one leg of R-212 must be clipped. You may regret doing this, however. See text for discusson of why.

"disable the beep." I can't imagine why anybody would want the "beep" disabled since it lets you know when you've hit or missed a key. This is very important when you're trying to work fast and accurately to program an armful of frequencies.

If you've worked at the keyboard of the PRO-2004/2005 at any length, you are well aware of how easy it is to miss a key. When you think you hit a key but don't hear a "beep" to confirm this, you instantly know to try again. If the "beep" is disabled, you will then be forced to constantly track your eyes between your list of frequencies and the LCD display to make certain that you haven't accidentally skipped over entering something. Have you ever programmed a consecutive, organized list of of 300 or 400 frequencies, only to discover towards the end that one at the beginning somehow didn't get programmed? This is especially relevant to the PRO-2004 with its keyboard as flat as Nebraska that offers little or no tactile sense when you hit or miss a key.

The "beep" was put there with good reason, but if you have a better one for ditching it, then don't let me stop you. It's easy to get rid of:

PRO-2004.

1. Disconnect the radio from its power source. Locate R-212 on the main receiver board in the left-rear quadrant, about a half-inch to the left of IC-3.

2. Clip one leg of R-212 part way down the lead so that it can be resoldered at some point in the future if you want the "beep" restored (and you will...).

3. You could also add an SPST switch somewhere and wire the two lugs of the switch to the two cut ends of R-212 to allow you to switch the "beep" feature on and off at will. That's probably the best approach because as soon as you disable the "beep" and begin to work with the mute scanner you'll wish you had the "beep" back.

PRO-2005.

1. Disconnect scanner from power source. Locate R-222 or C-219 on the main receiver board in the left-rear quadrant, about an inch to the left and forward of IC-3.

2. Remove either R-222 or C-219, but not both (it doesn't matter which one). These are surface-mount components, so be careful because someday in the future you will want to restore the "beep," possibly within 5 minutes.

3. You could also add a SPST switch somewhere and wire the two lugs of the switch as follows to switch the "beep" in and out of the circuit (see my comment for Step 3 of the PRO-2004 mod):

> A. Resolder one leg of the component removed in Step 2 to one of the solder pads from whence it came, but do it so that it stands on end.

B. Solder one wire from the new switch to the free end of the standing component.

C. Solder the other wire from the switch to the now unused solder pad where the component was removed in Step 2, above.

MOD-10   Making Scanners More Transportable

Sometimes it is most convenient, perhaps even necessary, to use a base station type scanner (such as a PRO-2004/2005) in places besides the radio desk at home. Most operators don't wish to buy (or can't afford) an extra scanner for every occasion and must adapt their main base scanner for any situation-- be the need in a motel, vehicle, temporary base, or some other weird place. One problem with this is the ungainly, awkward AC power cord that must be coiled up and then tucked out of the way when the unit travels or goes portable.

Some scanners (such as the PRO-2004/2005) were not particularly intended for mobile use. But why not, since they are capable of being powered from 12 VDC. No matter your application, this useful modification renders your scanner more easily transportable by making the AC power cord removable.

1. Cut the AC power cord about 2" to 3" from the rear of the chassis, the shorter the better-- but you have to leave enough to comfortably work with. And, of course, make certain that it has been unplugged from the power line before you do any cutting!

2. Strip about 1/4" of insulation from both wires at each cut end.

3. To the cut end of the short AC wire at the back of the scanner, attach a small power plug (Radio Shack #274-201).

4. To the cut end of the long AC power cable that will plug into the wall receptacle, attach a matching power socket (Radio Shack #274-202).

Safety Notice. This entire operation should be done with the power cord disconnected from the wall socket. Be sure to install the socket (female) on the end of the long AC cord that plugs into the wall-- **NOT** on the short wire stub on the rear of the scanner. The plug (male) goes on the short wire at the back of the scanner. If you it just the opposite, sooner or later you'll fumble around in the dark behind the scanner and grab ahold of those bare-- and electrically charged-- prongs for a severe and dangerous 117 volt shock.

5. If you operate your scanner in different locations, such as living room, office, den, basement, vacation home, as well as in the mobile or RV, you could make up several extra AC power cords with the mating socket and just leave them everywhere you operate your scanner. Make up another extra for the glovebox or trunk of your car so you can plug into an AC outlet at any location away from your usual haunts.

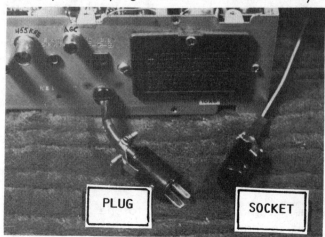

By adding this quick-disconnect 110 VAC power cord to your base station scanner, you can take it from place to place without dragging along the scanner's cumbersome attached cord.

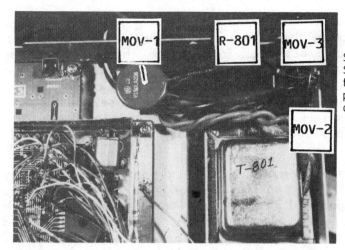

Spike and surge protection for base scanners. Safety is important. Insulate all exposed leads of the MOV's. Always unplug your scanner from its power source before removing it from its case and doing any work inside; 110 VAC can kill.

MOD-11 **Protecting Your PRO-2004/2005 from Voltage Spikes & Surges Also Other Scanners & Electronic Equipment**

Please realize that your modern keyboard programmable scanner is a very sophisticated package of computerized electronics. Maybe you take its high technology for granted, and you don't stop to think that it is comprised of thousands of transistors and other components that are set up so that if even one goes awry, all the rest of the components will react so negatively that the entire piece of equipment could stop functioning. Did I say "thousands of transistors"? Yes, some of those chips in your scanner each contain hundreds of transistors.

Getting off the track for a minute, the Allied Forces maybe could have won World War II in half the time had there been a PRO-2004 or PRO-2005 in the early 1940's. And yet few who own modern scanners ever stop to think that only a few decades ago, such equipment was not even considered a remote possibility using all of the devices, components, resources, and engineering talent that could be summoned.

And yet, if lightning strikes just a block away, a voltage surge could easily be introduced into the power lines. It could come sizzling down the power lines, right into the innards of your electronic MegaHertz inhaler, and have a hearty lunch on every transistor in sight. Unless you are willing to go out and buy a replacement scanner, you might just as well have found yourself out of the scanning hobby while also be the proud owner of $420 worth of newly-made large paperweight.

The frying of our scanners by voltages surges caused by lightning is to be deplored and discouraged. Same with damage caused by transmission line disturbances, the arc welding machine your neighbor owns, or even electromagnetic pulse (EMP) caused by high altitude detonations of nuclear weapons.

Power utility companies will normally supply clean, stable, constant electrical power through their lines, but from their earliest days they have disclaimed any liability for what their power does to things. Oil companies did that, too. I mean, when a terrorist tosses a Molotov Cocktail, the victims can't sue the oil company because they made the gasoline. And the farmer can't get sued because he grew the cotton that was made into the cocktail's wick. You've really got to look out for yourself these days and while you get what you pay for, sometimes you get a bit more.

That's why it pays to shell out a few bucks to protect your sophisticated electronics from those little extras that the power utility can't always control and for which it isn't willingly going to accept the blame-- like high voltage spikes that sneak along AC power lines looking for transistorized snacks.

Voltage spikes are caused by many different things, including power companies. Rarely can you ever prove the source of the single one that did in your equipment. Every time the washing machine's motor activates, for instance, there's a brief drop in the AC voltage throughout your home. The drop is accompanied by shorter, but sudden definite changes, high and low, of the AC power wave. These changes are in the form of "spikes" of voltage that (under certain conditions) can be several thousand volts. The time duration of these spikes is measured in microseconds (millionths of a second) and most of the time, because of the short duration, they are relatively harmless.

Sometimes, though, the right electrical conditions exist and something just quietly goes on the blink. A TV set that worked fine on Tuesday, but when you turn it on the following day, nothing happens. Strange. And stranger still that manufacturers of TV sets don't design surge protection into their products. Buy a sophisticated PRO-2004/2005, very susceptible to damage from voltage surges, and you' assume that the manufacturer would provide elementary protection. Not so! Your expensive scanner contains only very basic RF noise filtering where the AC power line voltage feeds the DC power supply, and this is most ineffectual against anything but RF, and one sometimes even wonders about that!

Voltage spikes in AC power lines can be caused by distant transmission line switching, maybe even a state or two away where the power is generated. Voltage spikes can be generated by nearby or distant lightning strikes; arcing transformers, washing machines, dishwashers, industrial equipment, and a host of other things. One interesting phenomena that is enough to scare the hell out of you concerns Electromagnetic Pulse (EMP). It has been postulated that a single, high altitude nuclear weapon detonated 150-miles above Kansas has the potential to damage or destroy communications and electrically powered equipment throughout the continental United States. How? Not by blast, heat, or radiation!

Rather, by the EMP that is generated at the instant of nuclear fission! EMP is a short time duration, wideband pulse wave containing radio frequencies from a few Hz to around 100 MHz. The energy content of this wave is extremely high and it travels at the speed of light, attenuating very little with distance. As the wave cuts through or across metal conductors (like power lines, AC power cords, antennas, fences, guy wires, coaxial cables, speaker wires, etc., it induces extremely high voltages in those conductors similar to lightning surges. Solid state equipment cannot withstand such an onslaught and is destroyed. EMP and its awesome destructive efforts have long been studied by the government, military, and electrical engineers.

Because EMP resembles a lighning surge and other more common voltage spikes, it can be defended against. In fact, much of our nation's nuclear testing program since the 1950's has been oriented towards <u>defense</u> against nuclear weapons on the theory that unless you blow some up you can't figure out how to defend against them. Two effective defenses against EMP and all kinds of voltage spikes are known to be effective. They are easily obtained and worth having.

The following instructions for voltage and surge protection of scanners were designed to be used with the Realistic PRO-2004/2005, however the techniques and components specified are readily adaptable to any electronic equipment that operates from 110 to 120 VAC. The diagram is Figure 4-11-1. Refer to that diagram and follow these instructions:

1. Procure three metal oxide varistors (MOV's) such as Radio Shack 276-568. MOV's are made by a number of manufacturers. General Electric's #V130LA208 is emminently suitable for this procedure.

Generally, the MOV should activate at about 130 VAC and be rated for 20 joules of energy dissipation, or greater.

2. Disconnect the scanner from the power source. Examine and become familiar

Figure 4-11-1
Voltage Protection

Typical Installation

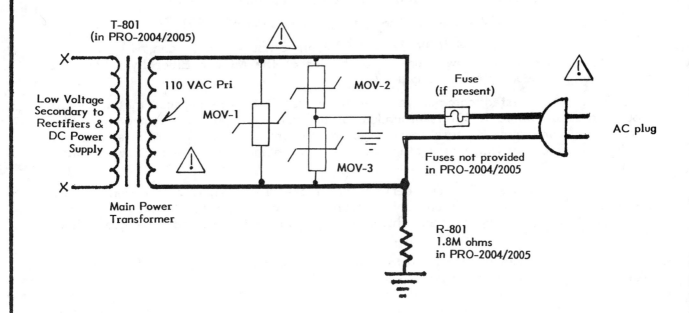

Notes - (See Text)

1. This procedure is functional for all radio and consumer electronics equipment if powered by a step-down power transformer and fed by 110 VAC @ 60 Hz.

    A. Use 3 metal oxide varistors (MOV) as shown (Radio Shack #276-568 or General Electric V130LA20B).

    B. Install MOV's as shown, but don't modify the AC circuitry in any other manner.

    C. **Important! Remove AC plug from wall receptacle before starting work!**

2. For maximum safety and protection, install a ground wire from the metal case or chassis to nearest true earth ground. Use ground rods or cold water pipes or AC electrical ground. Ground rods are best.

3. **Always be careful working with dangerous AC voltages. Never have any circuit or equipment plugged in to AC while you are working on it unless you are specifically instructed to do so for a brief test. <u>ABC= Always Be Careful!!</u>**

with the inside chassis area where the 110 VAC power cord enters the scanner and connects to the power transformer (T-801) in the PRO-2004 or PRO-2005.

You will see that two AC wires go directly to the primary leads of the power transformer, T-801. On one of the transformer primary leads, there is a 1.8 Megohm registor (R-801) that is soldered to a nearby ground on the metal chassis.

3. Clip both leads of MOV #1 as sort as possible, but so that it will still fit to the two primary leads of T-801 where the AC wires connect. Solder the MOV #1 to the primary leads of transformer T-801.

Note: It is important that the two leads on MOV #1 be as short as possible, and not bent, kinked, or curled, or coiled.

4. Solder one lead of MOV #2 to one of the primary leads of T-801.

5. Solder one lead of MOV #3 to the other primary lead of T-801.

6. Approximately position MOV #2 and #3 from Steps 4 and 5, and twist together the two free leads all the way up to the bodies of the MOV's. Liberally solder this twisted common lead of MOV #2 and #3.

7. Route the twisted common lead of MOV #2/#3 to the nearest metal chassis ground, possibly where R-801 is grounded, or a similar chassis ground spot and solder it to that ground spot. Be sure to first scrape the metal at the potential solder spot to clear away oxide, finish, and other surface contaminants that would prevent solder from adhering.

8. Inspect your work to make certain that there are no short-circuits and unsafe conditions. The exposed leads and solder connections of all the MOV's should be taped or covered with sleeving or heat shrink tubing. This will apply to all places where there are bare metal AC connections.

9. Button the radio up and operate with assurance that you have a digh degree of protection against voltage surges and transients.

A Touch of Theory. A metal oxide varistor (MOV) is a device that does absolutely nothing until its rated voltage is exceeded. The instant a voltage surge or spike comes in, the MOV senses it and turns into a short circuit that shunts the dangerous voltage away from the equipment being protected. When the dangerous voltage has passed, the MOV stops being a short circuit and returns to its normally passive status. Three MOV's are required to fully protect against voltage surges associated with power lines. One MOV hooked up as MOV #1 above offers substantial (about 80%) protection, but two more as with the case of MOV's #2 and #3 are required for maximum protection. The exact same hookup applies to all household electronic equipment powered by 110 to 120 VAC.

The theory here is that the transformer's primary is bypassed or shunted with MOV #1 to absorb any voltage surge impressed across the transformer winding. MOV #2 will shunt any voltage surge on one AC line to chassis ground, while MOV #3 will do the same for any voltage surge on the other AC line.

Other Forms/Types of Surge/Spike/Transient Protection. MOV's are one of the surest forms of protecting home electronic equipment from voltage surges in the AC power lines. One reason for this is because the MOV's are placed inside the equipment. There are other types of protection, some good and some not, which are briefly discussed as follows:

1. The most common variety of spike and voltage protectors are the kind you plug into a 110 VAC wall outlet. You then plug your equipment and appliances into one or more sockets provided on the protector. Radio Shack offers a number of these, including 26-203, 26-1365, 26-1395, 26-1396, 61-2187, 61-2780, 61-2781, 61-2786, 61-2791, 61-2792, and 61-2793.

The wall socket/multioutlet type of spike guard is better than nothing at all and is simple to install and operate. There is a weakness, however, in that a lightning strike or EMP can induce a high voltage in the AC line-cord between the protector and the equipment being protected. In such an instance, the protector is "upstream" from the problem and the equipment and therefore probably ineffective. Still, it will probably be effective for spikes attempting to arrive via the power mains. Therefore, it is much better than nothing at all.

2. Some operators think that fuses offer adequate protection against voltage surges. No way! Fuses and circuit breakers are "heat dissipation" devices. This means that they perform their intended function only when heat builds up to a certain level. At that point, fuses belt and circuit breakers pop. The heat buildup that causes them to do this takes time, which is just what your sophisticated electronic equipment hasn't got a lot of when it comes to dealing with a voltage surge. A typical voltage spike with potential for damage lasts for less than one-millionth of a second, and then it is gone leaving destruction in its aftermath. A fuse or circuit breaker requires tens of thousands percent more time than that in which to operate. Even "quick acting" fuses and circuit breakers need more than a tenth of a second to operate. A tenth of a second equals 100,000 millionths of a second, so you can see that by the time even the fastest fuse or circuit breaker reacted it would be far past the time when it would be of any help at all. Fuses and circuit breakers are great for fire safety and for long interval electrical disturbances, but not for short term spikes and surges.

3. You can achieve a major protection against voltage surges and spikes by coiling up the excess AC line cords of your electronic equipment. Fasten the coil with a rubber band or wire-tie and leave it hanging out of sight, out of mind. Coiled wire adds inductance to the wire, and this will notably impede short duration spikes and transients.

4. A gas-tube shunt is a form of surge suppressor that operates in a manner similar to an MOV. However, these are brute-force devices and, in any event, they don't respond as quickly as MOV's. They're virtually worthless on spikes and surges of less than a millisecond in duration, which are the ones most often encountered.

5. Zener diodes can be useful as surge/spike protectors, and are often used as such. Zener diodes aren't as long-lived nor as forgiving as MOV's. Remember, MOV's reset to the passive state as soon as the surge passes. A zener diode can be damaged by the surge. Also. zener diodes don't react as quickly as MOV's.

Get the idea, now? If you value your communications and other electronic equipment, then you'll employ MOV's and multioutlet surge/spike protectors very liberally throught your home on your scanners, communications receivers, computer, TV's, VCR's, stereo, AM/FM table radios, test gear, etc. I have placed not less than 51 MOV's and four multioutlet surge protectors in strategic use in my shop area alone. At the very least, put the three MOV's in your scanner so that it has a chance of surviving to be used when you might need it most.

Caveats & Warnings. Practically nothing known to man can survive direct hits by lightning, or nuclear bonbs. All the protection in the world isn't going to do any good for a direct hit by either. Close misses pretty much fall into the same category, too, although a near miss by lightning can possibly be survived if your equipment is properly protected. The thing is that destructive energy can enter electronic equipment through avenues other than power cords, such as antenna cables, and microphone cables, external speaker wires, tape recording cables, and through just about anything else that can be connected to your radio. Therefore, when you want the absolute utmost in protection, the only way is to disconnect everything from the radio and store it away in a grounded, shielded, metal container. In any event, the MOV modification described above offers a high degree of protection <u>only</u> from voltage surges that enter through the AC power cords. Of course, surges enter that way more frequently that via any other route, so it's a good starting point.

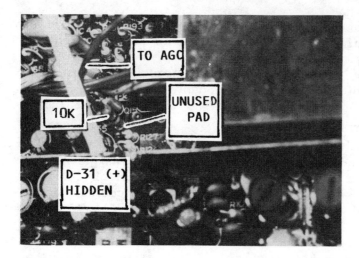

"S" meter circuit board wiring point for the PRO-2004. The best tap for the AGC is located at an unused solder pad located just to the left and rear of R-127. Install the 10K resistor as shown.

MOD-12 Providing An "S" Meter for Your
PRI-2004 & PRO-2005
Approach #1

An "S" meter (signal strength meter) is a most useful addition to the scanner that doesn't otherwise have one, and most scanners don't. An "S" meter provides an indication of the relative strength of an incoming signal. If the signal is weak, the meter indicates a low number. If the signal is strong, the meter indicates a high number. And there is a range in between.

For some reason, possibly as an economy measure, scanners have not come through with these devices. The ICOM R-7000 and the Yaesu FRG-9600 (which really aren't to be classed as scanners) have them, and I think that perhaps the old JIL SX-400 scanner had one. But, they don't make scanners with them, and that includes the PRO-2004/2005. It's relatively simple to provide these radios with an "S" meter, provided you already have or can obtain a digital voltmeter or a good quality analog voltmeter.

The method provided below provides a jack on the rear of the radio, which is connected to the scanner's Automatic Gain Control (AGC) circuits. A modern receiver has an AGC function that can be accessed easily enough. The purpose of the AGC is to develop a voltage that proportional to the strength of the incoming signal. What more could you ask for? The AGC is used at strategic places inside the scanner to control the gain. This modification will route a sample of the AGC signal to a jack on the rear of the radio, and then you connect a voltmeter to the jack for a "quick and dirty" meter that gives you an indication of relative signal strength.

The AGC voltage with which we will be working in the PRO-2004/2005 runs in the range of +0.130 VDC at a zero signal level to -0.310 VDC at a maximum signal, for a total range of about 0.440 volts-- not much, but enough. A digital voltmeter is best to use because then there will be several hundred points of resolution by which to evaluate incoming signals. An analog meter, to be effective, should have a 1-volt scale with a liberal zero adjust. The modification is easy, needs three parts, and will not harm or interfere with the scanner whatsoever.

PRO-2004.

1. Disconnect scanner from its power source. Install an "RCA phono plug" (Radio Shack 274-346) on the rear chassis of the scanner.

2. Install a 0.1 uF ceramic disc capacitor (Radio Shack 272-135) from the center lug of the phono jack to the ground lug of the phono jack.

3. Locate D-31 on the main scanner board. It is in the approximate center of the

board, just inside the shielded compartment having a cover with 13 alignment holes. Of course, you'll first have to pry that cover off. D-31 is also located adjacent to "TP-5".

4. Clip off all but 1/8" of one lead of a 10K resistor (Radio Shack 271-034). Tin that lead and solder it to the anode (+) of D-31.

5. Clip off all but about 1/8" of the other lead of the 10K resistor and solder a length of hookup wire to it. Cover that solder joint (and the resistor, if you like) with heat shrink tubing or other insulating sleeving. Tuck the resistor down out of the way so that it cannot interfere later with the shield cover.

6. Run the wire just soldered to the 10K resistor through one of the holes in the shield cover that you removed in Step 3. Replace the cover over the compartment from whence it came.

Note: Instead of running the wire through one of the holes in the metal cover, you may wish to cut a slot through the lip and top edge of the cover for wires to pass through. That way, the cover can be removed without the wires hindering things. There are other modifications that you might want to do which will require wires into and out of this compartment.

7. Solder the free end of the hookup wire to the new "RCA phono jack" installed on the rear in Step 1. The modification is completed.

PRO-2005.

Same procedures for PRO-2004 apply, except as follows--

1. Step 3: Instead of D-31, locate D-33, which is located in the left-center area of the main receiver board, just to the left of T-6.

2. Step 4: Clip off all but about 1/8" of one lead of a 10K resistor. Tin that lead and solder it to the anode (+) of D-33.

3. Steps 5 and 6: There may or may not be a shield cover to deal with.

Theory of Operation. The 10K resistor samples the AGC voltage developed by D-31. The resistor also isolates the receiver circuitry from anything that can happen outside of the scanner. You could, for instance, accidentally short-circuit the phono jack terminals and nothing bad will happen. The capacitor acts to filter out electrical noise and trash that might enter the radio via the new phono jack. It also helps stabilize any erratic AGC metering voltages and give the operator a visual indication of relative signal strength.

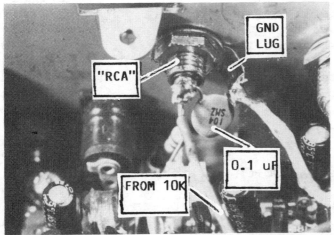

"S" meter chassis wiring for the PRO-2004 requires an "RCA jack" plus a 0.1 uF capacitor, some wire, and a ground lug.

The voltmeter will indicate slightly positive with +0.130 volts when no signals are present. It will swing negative to -0.31 volts when an exceptionally strong signal is present. Digital voltmeters have no difficulty with either polarity. An analog voltmeters must be connected with the red (+) probe to chassis ground of the RCA jack and the black (-) probe to the center contact of the phono jack. The analog meter should then be recalibrated to "0" when no signals are present. Signal activity thereafter will produce upscale meter indications.

MOD-13 Building an "S" Meter Circuit for Your PRO-2004 and PRO-2005 Approach #2

There are several methods of deriving an "S" meter for your scanner, most no better (and some worse) than Approach #1 in MOD-12. However, here is an interesting approach using a "receiver on a chip" and a small handful of components. The new IC has an on-board "S" meter function, so you couldn't hope for more.

I haven't tried this circuit, but it appears logical and sound, so it's included here for the hardier and more adventuresome scanner hackers. Actually, there's nothing to intimidate except that you're going to have to look beyond the convenient portals of the nearest Radio Shack in order to locate the IC and the meter. You'll have to get them at an electronics distributor.

This approach features internal circuitry that can't affect anything else, and a jack is provided on the rear chassis of the radio so an external meter can be connected. The neat thing here is that this circuit appears to offer a much wider voltage range than Approach #1 in MOD-12, so it should lend itself itself to your using a real "S" meter salvaged from a junked CB radio!

Table 4-13-1

PARTS LIST FOR S-METER (Approach #2)

CKT SYMBOL	DESCRIPTION	RADIO SHACK #
R-1	1.5-k ohms, ½-watt	271-025
R-2	150-ohms; ½-watt	271-013
R-3,4	10-k ohms; ¼-watt	271-034
VR-1	5-k ohm trimmer potentiometer	271-217
C-1	6-50pF trimmer capacitor	272-1340
C-2,5	.001-uF ceramic disk capacitor	272-126
C-3,4,8	.01-uF; ceramic disk capacitor	272-131
C-6,7	4.7-uF; electrolytic capacitor	272-1024
C-9,10	10-uF; electrolytic capacitor	272-1025
IC-1	HA1197, ECG1214 or NTE1214 integrated circuit	Note 1
IC-2	7805 Voltage Regulator	276-1770
J-1	RCA phono jack	274-346
P-1	RCA phono plug	274-339
M-1	analog meter movement, 200-uA full scale	Note 2

NOTES: 1) Obtain IC-1 and M-1 at local electronics distributor
2) M-1 can be a "real S-meter" salvaged from an old CB radio

Figure 4-13-1
"S" Meter for PRO-2004/5

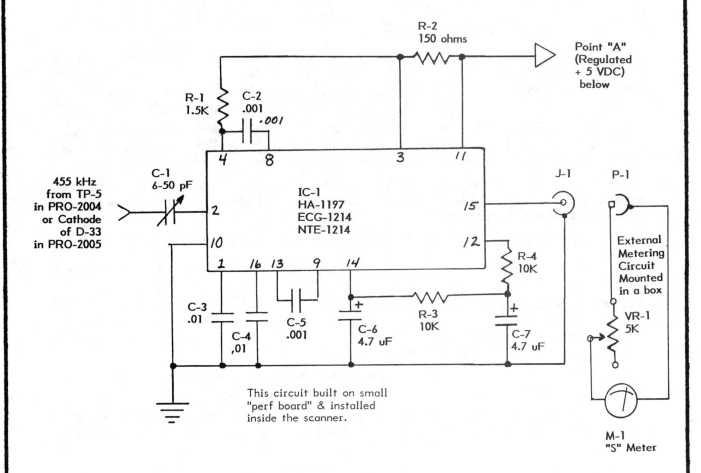

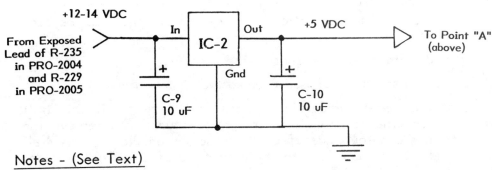

Notes - (See Text)

1. This circuit for scanners with 455 kHz last I.F. stage.
2. The DC voltage output at J-1 is proportional to the strength of received signals. Any DC voltmeter can be used in lieu of M-1, VR-1 & P-1 external metering circuit.
3. Input voltage (455 kHz) is critical. A 4.7 pF capacitor may be required in series with C-1.
4. Adjust VR-1 for maximum meter indication desired with a very strong signal received.

PRO-2004.

1. Review Figure 4-13-1, which is the wiring diagram. Also Table 4-13-1, which is the list of parts required for the project. Construct the circuit on a piece of "perf board" or design your own printed circuit board. Follow the wiring diagram explicitly, and do not question the pinout arrangement arrangement of the IC chip. Certain pins are wired unconventionally on purpose.

2. Disconnect the scanner from its power source. Physically install the "S" meter board inside the radio in the location of your choice. Be sure to connect the ground strip on the board to metal chassis ground of the radio or a circuit board ground point.

3. Connect Point "A" with hookup wire to the exposed leg of R-235. This supplies +14 VDC to the voltage regulator on the "S" meter board. R-235 is located in the left center area of the main receiver board and just to the left of connector CN-5.

4. Pry off the metal cover that has 13 holes in its top. Locate TP-5, which is the cathode of D-31. TP-5 is located in the center area of the main receiver board, just inside the shielded compartment.

5. Solder one leg of C-1 (from the new circuit's Parts List) directly to TP-5.

6. Solder a length of hookup wire or mini-coax to the free leg of C-1 previously connected to TP-5.

Note: It is preferable to use mini-coax such as RG-174 or at least some small, shielded microphone cable as the "wire" called for in Steps 6 and 7. If you must, a hookup wire will do, but it should be as <u>short</u> as possible. If you use mini-coax, be sure to ground the shield at both ends.

7. Solder the free end of the wire connected in Step 6 to Pin 2 of IC-1 on the new "S" meter board.

8. Install your new (or salvaged) "S" meter in a neat metal or plasic case, along with a patch cable and plug to match J-1. Note that VR-1 goes inside the case of the box.

Helpful Hint: You can use a digital voltmeter or an analog voltmeter, but an "actual" "S" meter would be ideal. Try to beg one from a shop that repairs CB radios, or offer to buy one of the dozens of junked CB chassis they probably have in the back room or basement.

9. Plug the "S" meter cable into J-1 on the rear of the scanner.

10. Power-up the scanner and set to an unused channel (no signals present).

11. Adjust C-1 so that the meter indication is close to "0" (zero). If this isn't possible, then again disconnect the scanner from its power source and remove the "wire" (steps 6 and 7) from the output side of C-1, and solder one leg of a 4.7 pF ceramic disc capacitor (Radio Shack 272-120) to the output leg of C-1. Then resolder the "wire" to the free end of the 4.7 pF capacitor. This reduces the coupling capacitance of C-1, which even at the minimum position, might be too much.
Power-up again and readjust C-1 for a close-to-zero reading.

12. Set the scanner to receive a known strong station, preferably the stronger the better (but only one using NFM-- not WFM-- mode).   Adjust VR-1 in the meter case for a full-scale (maximum) meter reading.

13. With VR-1 adjusted for strong signals, you may have to go back and repeat Step 11 to re-zero the meter for no-signal conditions. This could necessitate a touching up of VR-1 as the adjustments could interact with one another. Two or three repeats should suffice.

PRO-2005 only.

The foregoing PRO-2004 procedure is valid for the PRO-2005 with the following exceptions:

1. Connect Point "A" with hookup wire to either leg of R-229. This point supplies +12 VDC to 14 VDC to the 7805 voltage regulator on the "S" meter board. R-229 is located in the front-left area of the main receiver board just in front of connector CN-8.

2. Solder one leg of C-1 (from the "S" meter Parts List) directly to the cathode of D-33.

3. Solder a length of hookup wire or mini-coax to the free leg of C-1 previously connected to D-33.

MOD-14   Interfacing a Communications Receiver to Your PRO-2004, PRO-2005, or Other Scanner

The PRO-2004/2005 scanners are fine in many, many respects, but simply stated, they come up shy of operating features when compared with what you get when you buy a shortwave communications receiver. If you're a person who spent some time using a communications receiver, you'll notice these shortcomings right away. I'll illustrate by offering an outline of some of the deficiencies noted in the 2004/2005 chassis:

1. It lacks a "fine tune" feature for centering off-frequency signals, or for otherwise precision tuning. The best you can get in the way of precision tuning on the PRO-2004/2005 is the 5 kHz step rate in the "search" mode. This will allow fine tuning to maybe 2.5 kHz, but certainly no closer than that! In the "program" mode, you can enter frequencies in certain bands for a precision of 2.5 kHz, but no closer. This is not a major liability for the casual listener, especially since nothing short of the ICOM IC-R-7000 and Yaesu FRG-9600 class of receivers offers the ability to fine-tune signals. But read-on, because ability to "fine tune" implies a degree of selectivity that the PRO-2004 and most other scanners don't have.

2. The PRO-2004 is not a highly selective receiver, with published specs of -50 dB @ +15 to 18 kHz. For the FM bands above 30 MHz, this isn't too terrible, which is why the inability to "fine tune" isn't important to the casual listener. If you can "search" or "program" to within ± 5 kHz, that is plenty close enough, and (for most signals) your ears will never know the difference. If you're a purist, or if you want to use your $400 scanner like a $2,000 lab instrument, then the inability to fine tune is strictly bush league.

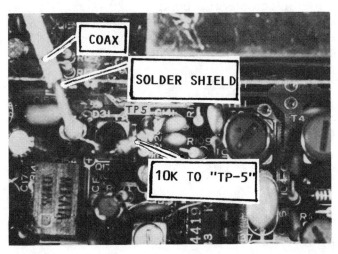

Shortwave interface wiring circuit wiring in the PRO-2004 calls for some mini-coax.

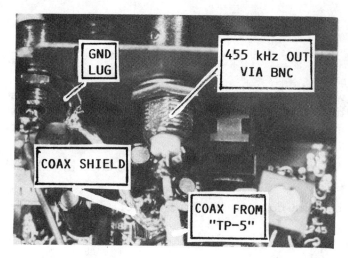

On the rear chassis of the PRO-2004, a BNC jack is used to feed the 455 kHz IF signal to the communications receiver.

Do you already, or would you like to use your PRO-2004 for frequencies below 30 MHz? Try scanning the CB or even the 10-meter Amateur band to see what I mean. Program in 27.025 MHz (CB Channel 6) and you'll also hear 27.015 and 27.035 MHz (Channels 5 and 7) at the same time. Not good, especially if you want to reliably use the "search" feature to keep up with the action on adjacent channels! One mitigating factor is that Realistic brand scanners have a nifty little circuit they call "Zeromatic" that activates in the "search" modes to seek the precise center frequencies before it locks on. "Zeromatic" doesn't do anything in the "scan" or "manual" modes, however.

The selectivity of the PRO-2004 chassis is determined largely by the scanner's 3rd IF filter, CF-2, which is a Murata-Erie CFW455D ceramic filter. It is especially designed to be wide-banded and not very selective because FM transmitter tolerances are for a bygone era. In other words, lots of the commercial equipment we monitor has rather sloppy frequency tolerance and even looser FM modulation specs. If the monitor receiver had a more selective IF filter, some of the stations you listen to wouldn't sound so good. There is a case, however, to be made for selectivity when tied to fine tuning capabilities for a receiver, and it's too bad that the PRO-2004 lacks in that department-- so far.

3. Now, here's a major liability of the PRO-2004: reception of single sideband (SSB) signals is virtually impossible. You say that there is no SSB activity on VHF and UHF, so why would a scanner need an SSB cabability? Well, CB channels (especially 27.365 to 27.405 MHz) have lots of SSB activity. You can also hear it on the ham bands (listen 50.20 MHz, 144.2 to 144.3 MHz, 903.1 MHz, and 1296.1 MHz). It's been noted at times in use for air/ground telephone calls in the 849 to 851, 894 to 896, 899 to 901, and 944 to 946 MHz bands. Also, industry has been pushing for ACSB (amplitude Compandored Single Sideband) for several years now. The FCC has recently authorized the ACSB mode for use on the new 1.2-meter commercial band at 220 MHz. It's only a matter of time before SSB becomes rather common in the scanner bands. FM is actually the least effective of the three common modes of voice communication (SSB, AM, and FM). SSB is the most effective. Watch for it to come.

Meanwhile, CB and the 28 and 50 MHz ham bands are filled with SSB signals that you can't presently make anything out of on your PRO-2004.

4. The PRO-2004 lacks other features that are considered standard by professionals and serious hobbyists. For the sake of brevity, these are itemized as follows:

    A. No "S" meter.
    B. No switchable Noise Blanker or Automatic Noise Limiter (ANL).
    C. No audio tone control.
    D. No user-selectable "selectivity" or bandwidth controls.
    E. No clock, timer or alarm functions.
    F. Many, many more too numerous to list here.

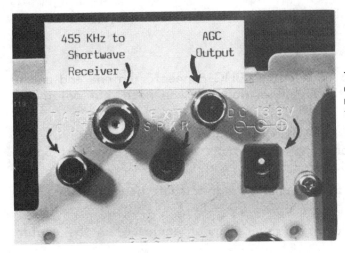

The added 455 kHz IF output jack is shown at the center-left in this view of the rear of the modified PRO-2004. The AGC output was added for the "S" meter.

But now, let's get on with adding some or all of the missing functions and features. All you need are a few common components, some basic handtools, a hank of miniature coax or shielded cable, a coax patch cable, and a communications receiver that can tune to 455 kHz.

Communications receiver? Well, yes. We are going to interface a shortwave communications receiver and all of its deluxe functions with the PRO-2004 to obtain the best of the two worlds for VHF/UHF scanning! The modification is very, very simple and it doesn't in any way negatively impact upon the performance, utility, or aesthetics of the PRO-2004. In fact, once the modification is performed, it is transparent to the user and will never get in the way or become a liability, even if it is rarely used.

Technically speaking, we are going to provide a raw, unprocessed IF signal from the scanner's 3rd IF stage (which operates at 455 kHz), to a jack on the rear of the chassis. A short coaxial cable will route this 455 kHz IF signal to your communications receiver which will take signal processing far past the point where the PRO-2004 stops. Provided that your communications receiver itself has the capabilities, you will realize the following "new" features for your VHF/UHF scanning capability:

1. User switchable wide/narrow selectivity.
2. Fine tuning cabability to ± 0 Hz, more or less.
3. SSB reception capability.
4. "S" meter for reading relative signal strength.
5. Clock, timer, and alarm functions.
6. Audio (sound) tone control.
7. Noise Blanker and Automatic Noise Limiter.
8. Maybe much more, depending upon the features of your communications receiver.

Much of the value of this modification is contingent upon your shortwave communications receiver's ability to receive FM signals. Some sets (like the Kenwood R-2000, for instance) have this capability, but if the set you have doesn't have this capability, then the mod may not be worth doing. On the other hand, if you scan or search the 25 to 30 MHz band to any extent, then it will be valuable regardless of whether or not your shortwave receiver has an FM capability. Most of the activity between 25 and 30 MHz is AM and SSB, with very little FM to be heard.

The modification is in two parts: (1) Perform simple internal work to the PRO-2004 or PRO-2005, and (2) Make a coaxial jumper cable with a BNC connector at one end on one end and a PL-259 plug on the other end. If you don't have some of the needed materials in your junk box, then the required parts might cost about $11 or so.

PRO-2005 Note: See the end of this modification procedure for exceptions and differences that pertain to the PRO-2005.

PRO-2004 Modification.

1. Make a jumper cable using RG-58 coax with a BNC connector on one end and a PL-259 plug on the other end. Make the cable as long as you need, but not much longer than necessary.

2. Disconnect the scanner from its power source, then remove the case of the PRO-2004. Drill a 3/8" hole in the rear chassis, just above or near the "Ext Spkr" jack. Install the BNC female chassis jack here.

3. Looking down on to the top of the scanner's main circuit board, locate Test Point 5 (TP-5) inside the PRO-2004.

Note: First, remove the top shield of the "square" inner enclosure in the center

Table 4-14-1

PARTS LIST FOR SHORTWAVE INTERFACE INSIDE THE SCANNER

QUAN	Description	Radio Shack #
1 ea	BNC female chassis jack,	278-105
1 ea	0.01-uF ceramic disk capacitor	272-131
1 ea	10-k ohm resistor, 1/4 watt;	271-1335
8"	mini coax or other shielded cable. Type RG-174 is preferred, but Radio Shack doesn't carry it. You can use Radio Shack's #278-752 if you can't beg or borrow some RG-174.	

Table 4-14-2

PARTS LIST FOR JUMPER CABLE

1 ea	UG-88 male BNC connector:	278-103
1 ea	PL-259 plug:	278-205
misc	Length of RG-58 coax cable: (whatever length you need to patch the PRO-2004 to your shortwave receiver, typically 2-4 ft.	278-1326

Figure 4-14-1
SW Receiver Interface
to The PRO-2004 & PRO-2005

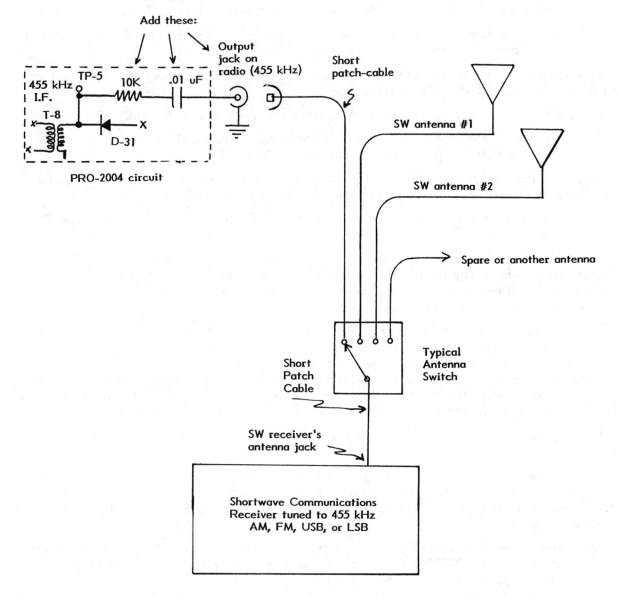

Notes - (See Text)

1. This interface allows reception of single sideband (SSB) and other signals picked up by the scanner. The shortwave communications receiver adds signal processing and enhancements that scanners don't have, such as: SSB, AM wide/narrow selectivity, fine tuning, noise limiter/noise blanking, "S" meter, audio tone control, and much more.

2. Both the scanner and shortwave receiver can still be used normally, too.

area of the scanner, the one with 13 holes in the top. There are only two inner enclosures, one is in the right-rear corner as you face the front of the scanner looking down. The other is in the center area of the scanner. Remove the "lid" or top of this center area enclosure by prying along the edges with a small screwdriver.

TP-5 is located near the exact center of the main printed circuit board as you look down from above. It is actually the cathode of D-31, but a special "post" that looks like an unbanded resistor is provided as TP-5 is just to the right of D-31.

4. Trim off all but about 3/16" of the lead from both ends of the 10K ohm resistor shown on our Parts List, and solder one end to TP-5.

5. Trim off all but about 3/16" of the leads on the 0.01 uF capacitor and solder one leg to the free leg of the 10K resistor.

6. Solder the center wire of the mini coax to the free end of the 0.01 uF capacitor.

7. Solder the shield of the mini coax to the nearby inner wall of the internal metal enclosure from which you pried off the lid in Step 3.

8. Solder the center lead of the other end of the mini coax to the center lug of the BNC female chassis jack which was installed in Step 2, above.

9. Solder or bolt the shield of the mini coax to the rear chassis near the BNC female chassis jack.

Note: A 3/8" i.d. solder lug for this purpose can be slipped over the BNC female chassis jack before tightening the nut. Or, you can wrap a solid wire around the threads of the BNC chassis jack before tightening the nut. Leave a bit of the wire hanging out to which the coax shield can be soldered.

10. Note where the mini coax comes out of the top of the inner metal enclosure. You can't replace the lid now, not until a slot is cut into the lid to accomodate the mini coax. Use a nibbling tool to cut away the edge of the lid to make an exit slot for the mini coax. If a nibbling tool isn't available, a diagonal-cutting pliers will work-- so will a pair of medium to heavy duty scissors, but watch that you don't cut a slot in your fingers in the process. The lid to be cut isn't very stout or thick.

11. Replace the lid of the inner metal enclosure, being careful not to pinch or cut the mini coax.

12. Temporarily energize the scanner and (carefully) perform basic tests to confirm that you've done everything correctly. Then, again disconnect the scanner from its power source as you replace the outer case and restore the PRO-2004 to its normal operating location.

13. Connect the BNC end of the new patch cable to the new BNC female connector on the rear of the PRO-2004. Connect the PL-259 plug end of the patch cable to the antenna port of your shortwave communications receiver.

14. Tune the shortwave receiver to 455 kHz, and tune the PRO-2004 to a VHF or UHF station (don't use an 88 to 108 MHz FM broadcast signal).

15. Vary the controls of the shortwave receiver to familiarize yourself with the newly found features and capabilities. Set the shortwave receiver to FM mode for normal scanner reception. For SSB reception, set the mode to either LSB or USB (as appropriate) and adjust the receiver tuning (clarify) for clear copy.

16. Vary the selectivity controls on the shortwave receiver to realize the benefits of fine tuning VHF and UHF signals from the PRO-2004.

Summary and Comments. Now it's done. You've just integrated two fine radios into one potent-performance package. What's more, you can still use each radio independently of the other, as desired. Yet, when you need them, you'll have the powerful features of both for your PRO-2004.

For simpler operation when you need the integrated package, it will be convenient to use a coaxial antenna switch to select among your shortwave antenna(s) or the PRO-2004. For example, you can connect one or more shortwave antennas to the switch, and connect a patch cable from the 2004 to a position on the switch just as if it were an antenna. Then, select any input you want, i.e., antennas or the 2004. Refer to Figure 4-14-1 for a typical interconnection diagram.

The output signal from the PRO-2004 will _always_ and _only_ be 455 kHz, plus or minus a few kHz (depending upon frequency errors in the transmitter you are monitoring). So your shortwave receiver _must_ always be tuned to or near 455 kHz while you're feeding in signals from the scanner. If the scanner is monitoring an FM signal (which will be for most VHF/UHF comms), the shortwave receiver must be set to FM. If the signal is AM or SSB, the shortwave receiver's mode should be set accordingly. Don't forget the aeronautical comms in the 118 to 137 MHz and 225 to 400 MHz bands are in AM mode.

Note: There is one mode of operation in the PRO-2004 where this neat little modification isn't going to work, and that's when the PRO-2004 is set to WFM (wide FM) mode, like when you're tuned to 88-108 MHz FM broadcasting stations. In this single instance, the PRO-2004 doesn't use the 455 kHz 3rd IF section. Instead, it has another circuit dedicated solely to WFM. No big deal, you probably don't have a lot of use for the benefits of a communications receiver while you're listening to an FM broadcasting station, anyway.

Differences to Note for The PRO-2005:

1. Wherever "TP-5" is mentioned for the PRO-2004, substitute "cathode (-) leg of D-33." There is no "TP-5" in the PRO-2005, but the cathode of D-33 performs the same function as D-31 in the PRO-2004.

2. The PRO-2005 employs different shields, compartments and covers from those discussed for the PRO-2004, but the principles remain the same.

Other Scanners. The process is simple. Determine the last I.F. frequency used by your scanner. Chances are that it will be 10.7 or 10.8 MHz. Identify and locate the "Detector" stage of your scanner, as it will be fed by this last IF frequency. Then, follow the safety, general connection, and interconnection instructions given for the PRO-2004/2005. Specifically, you will tap a sample (via a 10K resistor and a 0.01 uf capacitor in series) of the last IF stage's signal _before_ the detector stage, and route it to a jack on your scanner. Connect a coax patch cable between that jack and a shortwave receiver to 10.7 MHz (or whatever the IF frequency is for your scanner). Your shortwave receiver will now process that signal when you tune it to that IF frequency, adding its own unique capabilities.

MOD-15   **100 Extra Programmable Channels for The Realistic PRO-2004**

The PRO-2004 was a legend in its own time with its 300 programmable channels; no scanner with that many channels had ever streaked across the monitoring sky. But this fine scanner was hardly out on the market when a number of clever hackers simultaneously discovered that it was actually a 400 channel scanner in disguise. The microprocessor (CPU) chip was configured and programmed to address and process all 400 channels, but a block of 100 of those channels was omitted by the absence of a single critical diode.

That's right! All you need do is solder in a 10¢ silicon switching diode (like a 1N914 or 1N4148-- Radio Shack 276-1122) and your PRO-2004 will virtually become a PRO-2005! Your PRO-2004 will have 400 channels available in ten banks of 40-channels each and everything will function normally with no drawbacks or limitations whatsoever. The LCD display and the keyboard will function in all respects as if the 400 channels were designed in there by the factory (which they were)! Here's all you have to do to recapture those hidden 100 additional channels:

Want to increase the number of channels in your PRO-2004 from 300 to 400? Easy! See text to find & point the way to the D-510 slot where you will install an inexpensive diode.

1. Turn the scanner off and disconnect it from its power source.

2. Remove the case of the PRO-2004; locate the CPU compartment, which is PC-3 on the bottom side of the scanner. Pry off the metal cover of PC-3.

3. Find the row of diodes about 2" to the left of the 64-pin CPU chip. At the top of that row of diodes and just to the left of IC-504 will be two diodes marked "D-512" and "D-515". Between them will be two unmarked empty solder pads. These would be for D-513 and D-514. Just above D-512 are two more unmarked, empty solder pads. These are for D-510 and D-511, respectively, with the spot for D-510 at the very head of the row.

Observing the same polarity as all of the other diodes installed in that row, solder a 1N914 or 1N4148 diode to the spot for D-510. Make certain that the anode (+) goes to the left spot and the cathode (-) to the right spot.

Helpful Hint: It isn't necessary to remove PC-3 from the main chassis. D-510 can be soldered in from the top by placing a leg of the diode on the proper pad and then heating the pad with the soldering iron until the diode lead melts into the pad. Do the same for the other lead of the diode. It would help to presolder the two pads first before doing this.

That's it! Your PRO-2004 is now essentially a PRO-2005.

MOD-16 — 6,400 Programmable Channels for The Realistic PRO-2004 and PRO-2005 And Maybe Other Scanners, Too!

Attention: Realistic PRO-2004 and PRO-2005 (and maybe other) scanner owners. Have you already filled up your 300 or 400 programmable scan channels? Are your ten search bands filled to capacity? When you're snooping around in the search or direct modes, do the meager ten "monitor" channel slots cramp your style? Interested? Well, consider the following fantasy--

Programmable Memory Scan channels: 6,400 in 16 switchable blocs of 400 channels per bloc.

The 6,400 channel Extended Memory Board's wiring plan, bottom view. Construction will be the same for all sets, only the actual wiring points for the specific radio will vary. Note how the wires are brought through the "perf board" and bent before soldering to the pins of the chip. This prevents shorted pins and broken wires.

Programmable Search bands:	160, in 16 switchable blocs of 10 search bands per bloc.
Temporary storage channels for freqs. found in search/direct modes ("monitor" channels):	160, in 16 switchable blocs of 10 channels per bloc.

Fantasies can come true. This section shows you in detail how you can <u>add</u> 6,000 programmable channels, 150 programmable search bands, and 150 more monitor channels to your Realistic PRO-2004. Since the newer PRO-2005 is almost identical, from an electrical standpoint, to the PRO-2004, the modification will work for it, too.

Other Scanner Owners: This exciting modification can be adapted to other models of scanners that use an 8-bit static random access memory chip (SRAM) for data and channel storage. You could wind up with more or less than 6,400 channels, depending.

Rarely in the several decades I've been involved in communications engineering and technology have I been as enthusiastic about a retrofit performance modification as the one being presented here. During my years in communications, I've been involved in many facets, from Amateur and CB to satellites. Along the way, I have encountered or developed a fair number of exciting modifications to consumer electronics equipment, but putting 6,400 programmable memory channels into the PRO-2004 and PRO-2005 ranks up there among the best.

If you do the work and acquire the components yourself, and if you have an electronics junk box and a basic assortment of electronics hand tools and supplies, figure the cash outlay for this modification to be between about $25 and $50. Even if you start from near scratch with no tools or supplies, the cost could be less than $100. If you don't want to perform the "shirtsleeve" work, take this book and your PRO-2004/2005 to any communications shop that works on scanners and the technician there will probably do it for you. Of course, that will cost more, but the vast expanded memory should add an intrinsic value of $200-$300 to your scanner!

WHO can perform this memory modification? After having done it and painstakingly identifying all the required steps, I'll say that anybody who possesses basic electronic knowledge, basic soldering and hand tool skills, and a good measure of patience and thoroughness is capable of doing the job with success. The details in a step-by-step procedural format are given, and if you can follow directions and employ a bit of

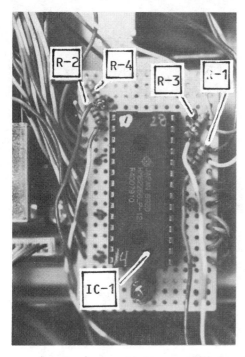
A top view of IC-1, wired in.

mechanical creativity, you'll be OK. Follow the directions explicitly and it's unlikely that any serious error or unpleasant surprises will occur. I made several potentially serious errors in the research/revision of this modification with no actual ill effects other than lost time. Decide for yourself what your capabilities and interests are. If you decide it's a "must have" modification, but is more ambitious than you want to attempt on your own, and there's nobody in your area you'll entrust the job to, all is not lost. See "If You Can't Do The Modifications Yourself" at the end of this chapter.

Who needs 6,400 channels and all the rest of the benefits of this modification? You do, if you're a serious scanner buff or if you are more than casually interested in your PRO-2004/2005. We all tend to make decisions based upon what we have to work with. If your scanner has 20 channels, then that's how many channels you'll make do with. If yours has 400 channels, then you'll live within those limits. When you have 6,400 channels, you'll probably never exhaust them all-- but your enjoyment and knowledge of scanning and monitoring will be profoundly expanded.

So, why would you want thousands of memory channels, scads of search banks, etc.? One reason would be to minimize the drudgery and effort that it takes to seriously pursue monitoring. The capacity to store all of the many configurations of scan channels and search bands instead of either forgetting or writing them down on scraps of paper that are soon lost is an example of what I mean. Also, consider that the monitoring spectrum of interest lies between 25 and 1300 MHz. That means that there are nearly 200,000 channels or frequencies of potential interest to you. There's no simple way to keep track on more than a hundred or so. This is one reason why scanner memory banks started out at 8, then went to 16 and 20, to 50, then to well over 200 in the past few years. In my own case, I use a computer to manage a data base of more than 2,500 frequencies used in San Diego County. When it came to scanning those frequencies, however, I had problems. Here's what I mean:

Imagine entering 400 channels at one time into your scanner; then monitor for a while; erase them; and repeat the entire cycle six more times. Now, imagine this process taking place regularly. Obviously, you'd hardly get any scanning done, because it it takes the better part of an hour just to enter and configure 400 channels, not to mention 2,500 or more! So, most of us "experts" confine our scanning activities to a select few essential channels that can be effectively monitored. We tend to leave out marginal frequencies because of the work and drudgery needed to get involved in the greater spectrum.

The 6,400 channel EMB, installed view. Note the cutout in the cover of the CPU compartment to permit the shield cover to be replaced.

Now, however, with 6,400 scan channels, I have everything I could possibly want right at my fingertips. Paperwork is reduced, I don't have to keep erasing channels to program in channels of interest I was missing, and I still have loads of empty channels to grow into. If you can't see the potentials for having a scanner that can deal with 6,400 channels, then you might wish to refresh your memory of Aesop's fable of the Fox and The Sour Grapes-- and then get down to this modification!

The PRO-2004 came from stock with 300 programmable channels, 10 search bands, and 10 monitor channels. As noted in MOD-15, enthusiasts soon realized that it could be expanded out to a 400 channel memory with the mere installation of a cheap switching diode. The newer PRO-2005 is practically a replica of the PRO-2004 with that same diode installed by the factory, plus some aesthetic redesign and a different model number. The electronics inside the PRO-2005 are physically different with a redesigned printed circuit board (PCB) and a greater profusion of Surface Mount Technology (SMT) components. Yet, in the final analysis, the electronic design of the PRO-2005 is practically identical to the PRO-2004. This section is written expressly for the PRO-2004, but notes and suggestions are made for the PRO-2005. The major differences between the two are only mechanical or physical, and not electrical.

If you have an **unmodified** PRO-2004/2005, you may want to perform some or all of the other modifications in this book before you perform MOD-16. Be sure to get a Service Manual (See Table 4-16-3). You must have the Service Manual both for reference as we go along as well as to assist you in debugging or resolving any problems after you've done the modification.

Credit & Acknowledgements. Credit and honorable mention most appropriately go to Terry Vaura, who apparently did the first version of this modification. David Jones, did some revisions to Mr. Vaura's version. Still, both versions lacked essential detail and were apparently intended for advanced hackers, technicians, and engineers. I have reviewed these earlier works, gone into greater detail, and added a few new twists and kinks of my own to make this memory modification more feasable for nearly everyone.

Theory of The Modification. The PRO-2004 employs a surface mount 2K x 8 Static Random Access Memory (SRAM) chip (16K) for storage of Programmable Channels, "Search" bands and "Monitor" channels, together with the various modes (AM, NFM,

The 6,400 channel EMB, view of wiring to PC-3.

WFM), "Search" increments (5, 12.5, and 50 kHz), and "Lockout" channels. The stock 24-pin memory chip has the circuit symbol "IC-504" and is either a TC5517CF-20 or uPD446G-45. The IC-504 is located inside the CPU module (PC-3) and will be removed and replaced by a 32K x 8 SRAM (256K) chip in a DIP package with 28 pins. The stock IC-504 is a surface mount type so there is no requirement or special need to remove the CPU module from the chassis. All work can be done after disconnecting the scanner from its power source, removing the main case of the scanner, and prying off the metal cover of PC-3. IC-504 will be replaced by the new 256K chip which will be installed on a small "perf board" and hard wired, practically pin-for-pin, to where IC-504 was removed. The perf board holds four resistors; one capacitor; the new memory chip; 28 wires, and will be mounted outside of and adjacent to the CPU module.

The static RAM chip and the CPU chip work together hand-in-hand to make the scanner function the way you know the unit. The CPU chip is unique and cannot be modified or exchanged, but the static memory chip is standard generic and will be substituted with one having far greater capacity. The stock 16K memory chip has 11 "address" pins which the CPU is designed to query and gate (A0-A10). The new memory chip has 15 address pins (A0-A14). The CPU will work as normal with the first 11 addresses of the new memory chip (400-ch., etc.). The CPU has preprogrammed ROM, which neither this nor any modification can alter, so external switching is required to access the other four address pins for 15 additional 400-channel memory blocs. A four-segment DIP switch will be mounted on the front or top panel of the PRO-2004 to permit external selection or switching of the additional four addresses to yield a total of 16 blocs of 400 channels each. In a nutshell, that's it. The mod is more time consuming than technical.

PRO-2005: The PRO-2005's SRAM memory chip has the circuit symbol "IC-505" and is identical otherwise in all respects to IC-504 in the PRO-2004. It may be either a TC5517CF-20 or an LC3517BM-15. The IC-505 is located on the PRO-2005's CPU board, also known as PC-3. I have not yet been "inside" a PRO-2005, so I can't say for sure how available IC-505 is for removal and replacement, but from the Service Manual, it appears just as accessable as in the PRO-2004. Most of the procedures, tips and kinks given for the PRO-2004 should apply or be adaptable to the PRO-2005. Specific comments are given where applicable.

The New Memory Chip. The replacement memory chip is specified to be a Hitachi HM62256LP-15. The suffix doesn't necessarily have to be -15 and can also be -12, -10, or even -20. In memory chips, the suffix after the dash indicates the speed of the chip, with lower numbers being faster and the higher numbers being slower. <u>Do not</u> use a memory chip with a suffix larger than 20. Another thing on which to take note is the "LP" just before the suffix. This means "**L**ow **P**ower" and is an important

The 6,400 channel EMB, overall view. There's still room available for additional modifications.

feature. Other memory chips without the low power feature will draw more current from the memory retention battery when the scanner is not plugged in to AC or DC power. Whichever brand of chip you select, be sure to specify "low power" and medium to fast speed. I have specified the Hitachi chip because <u>I know it will work</u>, but there are other SRAM 32K x 8 memory chips that should work just as well.

Finding a 32K x 8 SRAM chip isn't difficult, but you have to know where to look. Forget about Radio Shack, they do not stock the item-- but most solid state component distributors do. A few probable sources to check, if you have none local to you, are given in Table 4-16-1.

Here's how to order: By letter or phone, ask for Hitachi HM62256LP-15, -12, or -10. If they are out of stock, or if the price is seems too high, then ask for a quote on any "low Power SRAM. -10, -12, or -15 speed, and 256K memory organized as 32K by 8." Use those exact words, and you'll be OK. Expect to pay about $18 to $25.

Caution: This memory chip is fragile, meaning that it can be destroyed by the tiniest static charge. When you buy one, it will come imbedded in a special carbon foam, enclosed in a special plastic wrapper for static protection. Leave the chip in its packaging until you are ready to install it. We'll use a 28-pin IC socket, so minimal handling and no soldering of the chip will be required. Other precautions will be given in the Steps of Procedure.

Table 4-16-1

<u>SOURCES OF DIGITAL & SOLID STATE PARTS</u>

LA PAZ ELECTRONICS INTERNATIONAL
9400 ACTIVITY ROAD; SUITE "C"
SAN DIEGO, CA 92126

MICROPROCESSORS UNLIMITED, INC.
24000 S. PEONA AVENUE
BEGGS, OK 74421

JDR MICRODEVICES
2233 BRANHAM LANE
SAN JOSE, CA 95124

SOLID STATE SALES
PO BOX 74-D
SOMERVILLE, MA 02143

JAMECO ELECTRONICS
1355 SHOREWAY ROAD
BELMONT, CA 94002

DIGI-KEY CORPORATION
PO BOX 677
THIEF RIVER FALLS, MN 56701

Table 4-16-2

PARTS LIST
FOR EXTENDED MEMORY MODIFICATION

CKT SYMBOL	QUAN	DESCRIPTION/PART NUMBER
IC	1	Memory IC chip, 32k x 8 SRAM; Hitachi HM62256LP-12, -15
C-1	1	Capacitor, tantalum; 0.1uF/35 WVDC; Radio Shack #272-1432
R_{1-4}	4	Resistors; 4700 ohms, ¼-watt; Radio Shack #271-1330
S-1	1	DIP Switch; 4PST; Radio Shack #275-1304
J-1	1	IC Socket; 28-pin; wirewrap Radio Shack #276-1983
J-2	1	IC Socket; 8-pin; wirewrap Radio Shack #276-1988
MISC	1	Perf board; Radio Shack #276-1395
	20 ft	Wire; stranded; flexible; 22-ga; Radio Shack #278-1307
	1 pkg	Standoffs for circuit boards; Radio Shack #276-195

Table 4-16-3

TOOLS & MATERIALS LIST
FOR EXTENDED MEMORY MODIFICATION

QUAN	ITEM DESCRIPTION	COMMENTARY
1	Service Manual for PRO-2004 or PRO-2005	Obtainable from Tandy National Parts Center; 900 E. Northside Dr. Fort Worth, TX 76102 (817) 870-5600
1	Soldering pencil; 15-25 watts with small, conical or flat tip	Keep tip clean with wet sponge during work.
20 ft	Rosin core solder; .032" dia	Radio Shack #64-005
1 rl	Desoldering braid or wick	Radio Shack #64-2090
1	Rosin Flux remover; spray can	Radio Shack #64-2324
1	Electric drill & assorted bits, including one small bit of about .05"-.075" diameter	See text for alternatives to the small drill bit
1	"Xacto" knife and/or several single edged razor blades	
1	Sheet metal "nibbling tool"	Radio Shack #64-823
1 pk	Super glue; fast setting	Radio Shack #64-2308
Misc	Basic hand tools including phillips & flat blade screwdrivers; pliers; diagonal cutting pliers; jeweler's screwdrivers; magnifying lens; several sizes of sewing needles & straight pins; mechanic's metal file; hacksaw; pocket penlight with fresh batteries;	

Figure 4-16-1
Wiring Diagram: 6,400 Channel Memory Modification

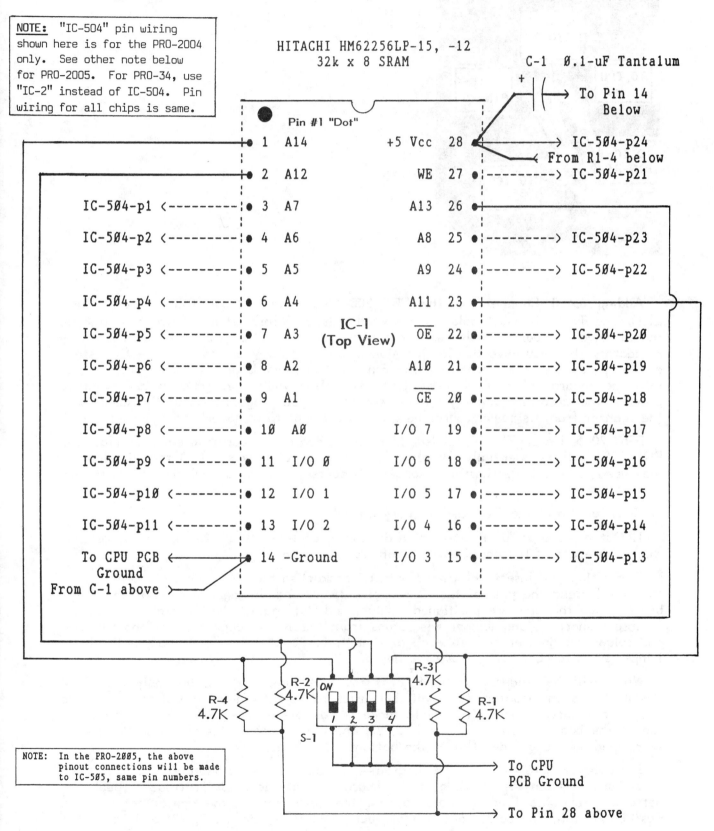

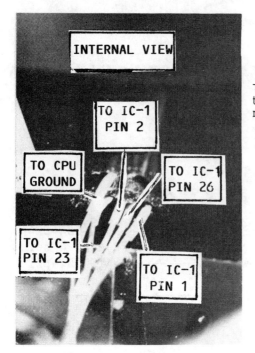

The 4-segment DIP socket installation, internal view. Take the front panel from the chassis and you have plenty of room in which to work!

Adding 6,400 Channels to The PRO-2004/2005. Review our Table 4-16-2 (Parts List), and Table 4-16-3 (Tools and Materials List). Before sitting down to work, you should have all parts, tools, and materials handily and organized. It is absolutely mandatory that you have the Service Manual on hand for your scanner. The first stage of the procedure is to construct the Extended Memory Board, so don't start to dig into the innards of your scanner just yet unless you're just curious and want to preview the areas in which you'll be working as you progress. Naturally, disconnect the scanner from its power source before you attempt to remove it from its case.

PRO-2005 Note. The Extended Memory Board as constructed here for the PRO-2004 will be essentially the same for the PRO-2005, except that the wire lengths may need to be a little shorter or longer. Of course, you can always trim them to size later on.

Construction of the Extended Memory Board:

1. Cut a piece of "perf board" to a dimension of 1-1/2" W by 2-1/4" L. File or sand the rough edges of the board until smooth to the touch.

2. Position and insert J-1 (the 28-pin IC socket) so that it is precisely centered on the board. Bend the pins of the IC socket outward and flat against the bottom of the board. Once the pins are positioned straight and flat against the bottom of the board, insert short 1"-long jumper wire stubs from the top through the perf board holes and solder one jumper stub to each of Pins 1, 2, 23, and 26. Trim the length of the jumper stubs to about 1/2" above the board.

Wires will be soldered to each of these four jumper stubs, but only after the module has been mounted, so it will be easier to work from the top of the case. Use your own creativity and initiative to make it possible to later solder a wire from the top of the board to each of Pins 1, 2, 23, and 26. All other pins of the socket should be clipped and dead-ended flat on the bottom side of the board.

3. Review Figure 4-16-1-- the diagram-- and accompanying photographs. Install the four resistors from the top side of the board so that the leads protrude through the bottom. Insert the four resistors so that they are flat on the board; not on end. Position, install, and solder one end of each resistor to Pins 1, 2, 23, and 26 on the

An external view of the 4-segment DIP switch, showing its installation. The author used a 16-pin socket for other purposes. You need only an 8-pin socket for this modification.

bottom side of the board where the jumper stubs were soldered in Step 2. The free (unsoldered) ends of the resistors should be pointed away from and not be close to any of the IC socket pins.

4. Solder the free ends of each of the four resistors to a bare wire of appropriate length. The bare wire may have to be routed slightly around the Pin 1 and 28 wnd of the board. Be certain at this point that it does not touch anything other than the four resistor leads.

5. Install capacitor (C-11) as follows: The (+) lead to Pin 28 of the IC socket. The (-) lead via a short splice (if necessary) to Pin 14 of the IC socket. Note: This capacitor may be installed on the bottom of the board, if desired, with the leads diagonally stretched between Pins 28 and 14 of the IC socket. It is <u>important</u> that the (+) lead of the capacitor be soldered to Pin 28 of the IC socket.

6. Solder a short jumper wire from Pin 28 of the IC socket to the common bare wire that was soldered to the free ends of the four resistors in Step 4.

<u>Check and recheck all your work up to this point. Assure perfection</u>. You might use a fine-tipped indelible marking pen at this time to mark the pin numbers of the IC socket on both the top and bottom sides of the board. This will make your later work somewhat easier with reduced chance of error.

7. Prepare 24 color-coded hookup wires as follows:

    A. Wire type: insulated, stranded, <u>very flexible</u>, about 22-24 ga.
    B. Length: about 8", but longer is OK.
    C. Strip about 1/8" of insulation from <u>one</u> end of each wire. <u>Do not</u> strip the other ends at this time since they will be positioned later and clipped for precision fit.
    D. Tin the stripped end of each wire with solder.

Note: You can use 25-conductor "ribbon" wire, but if you do, you'll soon lose track of where the wires go. Better off getting a foot of 25-cond computer or telephone cable where the wires are individually color-coded. Strip the cable and salvage the wires for this job.

8. Position the board so that Pins 1 and 28 are at the top, and Pins 14 and 15 are at the bottom. Insert a previously prepared wire from the <u>top</u> of the circuit board about 3/16" into the hole along the <u>outer</u> edge of the perf board next to Pin 3 of the IC socket (J-1). Turn the board over to view the bottom, and bend the tinned wire end 90° so that it touches the top of Pin 3 of the IC socket. Solder this tinned end of the wire to Pin 3 of the IC socket.

After the solder connection has cooled, pull the wire tight from the top of the board so that all slack is taken up from that portion of the wire on the bottom of the board. Ensure a smooth, solid solder connection of wire to Pin 3.

9. Repeat Step 8 for each of the remaining 23 wires. One at a time, solder the tinned end of each wire to a pin of the IC socket (J-1). <u>Do not</u> solder any wires to the jumper stubs at Pins 1, 2, 23, and 26 of the IC socket at this time. Jot down a record of the color code and pin numbers as you go-- you will need this for later reference.

Note: The IC socket pins are rather close together, and solder particles or just plain sloppy soldering can cause adjacent pins to short together. That happened in my prototype design and later caused needless frustration, troubleshootong, and time loss.

<u>Check all of your work again at this point.</u> Use a strong magnifying lens to scrutinize each solder connection. Look for shorts between adjacent pins and for poorly taken solder joints. Redo all questionable or defective solder joints at this time. Spray the bottom of the board with Rosin Flux Remover to clean away flux residue and to remove loose particles of solder. Again, inspect the bottom of the board with a strong magnifying lens. Finally, use a small-tipped tool such as a jewelers screwdriver to manually flex and prod each soldered pin of the IC socket. Resolder any that came loose. Verify again that there are still no shorts between any adjacent pins of the socket and that all wires, resistors and the capacitor are properly soldered. The Extended Board is now complete. Put it aside in a safe place.

Construction and Installation of the Dip Switch Interface - Pro-2004/2005. Four simple SPST on/off switches are used to access the memory of the new IC chip. The CPU will automatically address whichever of sixteen 400-channel bands is manually selected. While this procedure specifies a 4-segment DIP switch for manual selection, you are by no means limited to this means of control. Four separate SPST toggle switches can be used, for instance. I suppose even a thumbwheel switch could be employed as could any of a dozen other switching arrangements. The essential requirement is a separately selectable "on" and "off" status to each of Pins 1, 2, 23, and 26 of the Extended Memory chip.

Digital means can also be employed as proposed by Terry Vaura, who did a design using additional circuitry in the form of a binary counter chip, an EPROM and a seven-segment display LED. While in keeping with modern technology, this method makes the project much more cumbersome and puts it out of reach of the average hobbyist who isn't an engineer or advanced technician. Therefore, the procedure here employs a simple, but no less effective, switching arrangement. Again, the main thing is that Pins 1, 2, 23, and 26 of the chip need to be independently and cooperatively addressable in order to access the 16 memory blocs. Four switches with on/off positions allow 16 combinations of switching "on" and "off" which is required to address the full 6,400 channel capability of the memory chip. Basically, it's like this: one switch will control two groups of 400 channels each. Two switches control four blocs of 400 channels; three switches control eight blocs of 400, and four switches control the full 16 blocs of 400 channels!

The question is where to install the switches. Space is at a premium in both the 2004 and 2005. The rear chassis contains ample room, but is inconvenient to access during operation. The front panel offers possibilities, but I elected to use the top of the plastic escutcheon directly above the LCD display panel, centered over the "tic" portion of the word "Realistic." See the photo.

I suppose that you <u>could</u> build the switching scheme into a small metal box and route the 6-wire cable <u>bundle</u> into the scanner via a hole in the rear of the chassis. If you go this route, you're on your own because I have an aversion to lengthy wire runs where digital and RF signals are concerned. External wires are an invitation for noise pickup and all sorts of things could happen-- none of them good. Then, again, it just might work. Like I said, you're on your own if you choose that way.

At this time, disconnect the scanner from its power source, remove the four screws from the rear of the scanner and slip off the outer case. You may also remove four countersunk phillips screws from the sides of the front panel to permit it to swing down for easier inspection of the inside areas. <u>Decide upon your switching arrangement at this time</u> and then follow the below Steps of Procedure, or modify them to suit your needs.

PRO-2005 Note: The mechanical procedures described in Steps 10 to 23 are specific for the PRO-2004. If you find a different place for and still choose the 4-segment DIP switch for external control, you may use the electrical procedures verbatim. Wire lengths may have to be longer, etc., but the following procedure will guide you well enough.

10. The AC plug is disconnected from the wall. Remove the four screws from the rear of the chassis. Slip off the cover. Remove the four countersunk phillips screws that hold the front control panel to the chassis; two are on each side of the front panel. Allow the front panel to slightly tilt away from the chassis.

11. Cut a small piece of "perf board" for a drill template as follows:

   A. Dimensions to be at least 6 holes wide by 10 holes long.
   B. Temporarily position J-2 (the 8-pin IC socket) on to the template board so that its leads protrude through 8 of the holes. Darken those holes with a marking pen.

12. Position the template fabricated above on to the top of the front panel, just above the LCD display and centered over the letters "tic" in the "Realistic" logo. Tape the template down flush to the top of the plastic case.

13. Drill 8 holes down through the marked template holes, through the top of the plastic front panel. This calls for a very small drill of about .05" to .075" diameter. The perf board holes are .042" dia., but we want the holes through the plastic case to be a bit larger than that.

There is a suitable alternative to drilling. Using the template made in Step 11, <u>melt</u> holes through the plastic case with a heated sewing needle or a nail of appropriate diameter. Yes, this will make a bit of a mess at first, but is easily cleaned up. The idea is to drill or melt 8 holes into the top of the plastic front panel so that an 8-pin IC DIP socket can be mounted there. The holes need to be larger than the socket pins to accomodate the wires that will later be soldered to each pin of the socket.

After drilling or melting the holes to proper size, <u>carefully</u> use a single-edged razor blade to slice or shave off the distorted plastic "mounds" so that the surface is flush and smooth as before. Watch your fingers-- the blades are sharp.

14. Prepare four color-coded hookup wires as follows:

   A. Wire type: Insulated, stranded, <u>very flexible</u>, about 22-24 ga.
   B. Length: About 8" to 10".
   C. Strip about 1/8" of insulation from <u>one</u> end of each wire. <u>Do not</u> strip the the other ends at this time, since they will be positioned and clipped to fit later.
   D. Tin the stripped end of each of the four wires and solder.

16. Carefully solder the four wires as prepared in Step 15 to four pins on one side of J-2. These four pins must all be on the <u>same side</u> of the IC socket (J-2), one wire to a pin.

17. Prepare 4 more wires as follows (color code not necessary):

   A. Wire type: Insulated, stranded, <u>very flexible</u>, about 22-24 ga.
   B. Length: About 2".

    C. Strip about 1/8" of insulation from <u>one</u> end of each wire and about 1/4" from the other end of each wire.
    D. Tin the stripped 1/8" ends with solder, but not the other ends.

18. Solder the 1/8" strip and tinned wire ends to the remaining pins of J-2, the 8-pin IC socket. These four wires must all be on the <u>same side</u> of the IC socket, and <u>opposite</u> those soldered in Step 16.

19. Lower the prewired J-2 (8-pin IC socket) into the 8 holes drilled or melted into the top of the front panel in Step 13. Push the IC socket down slightly until it seats. Check for a perfect fit and good appearance. Make adjustments using a drill, heated needle, "X-ACTO" knife or single-edged razor blade until the fit is perfect. Use care when working with the sharp blades.

20. Slightly raise the prefitted J-2 and apply a <u>small</u> bead of "super glue" type adhesive around the underside of J-2, but don't let the glue get on the pins (or your fingers). Firmly reseat J-2 down flush with the top of the front panel and apply pressure until the glue sets up and bonds.

21. Twist together the 1/4" bare ends of the four short wires from J-2 installed in Steps 17 and 18. Solder these four wires together into a single joint. Splice and solder a single 20-24 ga. stranded, insulated wire about 3" to 6" long to this 4-wire junction. Insulate this soldered connection with tape or heat-shrink tubing.

22. Reinstall the front panel to the chassis with the 4 countersunk phillips screws removed in Step 10. Tighten securely.

23. Solder the free end of the wire fashioned in Step 21 to any printed circuit board grounding spot. Do this anywhere on PC-3 (the CPU module). All of the other 4 separate wires from J-2 to hang free and accessible.

This completes most of the installation of the switch interface.

Removing the Stock Memory Chip (PRO-2004): With the power plug inserted, turn the scanner on to ensure that nothing is amiss and there are no problems so far. The scanner should be functioning normally in all respects. Before turning the scanner off, remove the Memory Retention Battery from its clip on the rear of the radio. Turn the scanner off, and with the "Restart" switch (on the rear of the scanner) depressed, turn the set on again. Memory should be cleared now and the battery indicator will flash with an occasional "beep" sound. Turn the radio off and <u>unplug it from all AC and DC power</u>. Pry off the metal cover of the CPU compartment, PC-3.

24. Locate the stock memory chip, IC-504, in the CPU module.

PRO-2005 Note: In the PRO-2005, IC-50<u>5</u> will be removed, <u>not</u> IC-504. The following removal and cleanup techniques are applicable.

25. Remove IC-504 (PRO-2004) using desoldering braid and/or desoldering suction devices. The desoldering braid/wick is probably the best approach to this particular situation. It will take several heatings and reheatings before the braid has drawn up all the solder from the pins of IC-504. Even then, IC-504 will remain firmly attached. Gently pry each pin upward with a sewing needle while the pin is reheated. Maintain an upward pressure until the pin pops free. <u>Be very careful here</u> to leave the solder pads for IC-504 intact and damage-free. Soon, IC-504 will pop free and come loose. Store the chip in the same packaging materials as your new memory chip is stored.

26. Use more desoldering wick to clean up the solder pads from where IC-504 was removed. Make them bright, shiny and clean. Spray the cleaned pads with Rosin Flux Remover to get rid of any residue.

Mechanical Installation of The Extended Memory Board (PRO-2004): The Extended Memory Board should be mounted before wiring is attempted. It is my understanding that some have installed their Extended Memory Boards right inside of the CPU compartment where IC-504 used to be, which is perhaps ideal from a total purists point of view, but actually it isn't necessary. This is fortunate, because the wiring inside such cramped quarters would be a nightmare for the uninitiated. For the majority of us, a spot outside of and adjacent to the CPU compartment is better, for we then have lots of room in which to work.

With the scanner facing you, but upside down, inspect the area of the inner metal chassis just to the right of the CPU compartment and just behind the front panel numerical keypad. You will see a couple of square inches of exposed metal chassis-- which is an ideal mounting spot for the Extended Memory Board. You'll need one standoff or tubular bushing of a length of about 1/8" to 1/2" to hold the Memory Board rigidly in place and yet up off the metal chassis. You could use a machine screw and several nuts as necessary to comprise the spacer.

PRO-2005 Notes: You are more or less on your own in Step 27, but the theory and principles given for the PRO-2004 are valid enough for the PRO-2005. The differences are mechanical and physical and you'll have to adapt the installation techniques to what best suits the PRO-2005.

27. See the accompanying photos and inspect <u>both</u> sides of the inner metal chassis, just to the right of the CPU compartment and just behind the front panel numerical keypad. There are some components mounted on this inner metal chassis on the side opposite the CPU compartment-- you certainly will <u>not</u> want to drill a hole through them! There is plenty of room to drill, however, if you look around. Select a spot for your standoff, and drill a 1/8" hole. See the photos.

28. Drill a 1/8" hole in the center-end area of the Pin 14/15 end of the Extended Memory Board.

29. Insert a machine screw of appropriate length through the newly drilled hole in the chassis so that the screw protrudes up through the CPU side of the chassis. Slip the standoff or several nuts on to this machind-screw. Tighten securely. Position the Extended Memory Board on to the machine-screw and attach another screw or nut. Tighten.

This completes the mechanical installation of the Extended Memory Board.

Electrical Installation of The Extended Memory Board (PRO-2004): Carefully study the solder pads from where IC-504 was removed. You'll note that they are exceptionally close together. It is fortunate, however, that there are <u>other</u> places where we may solder the wires from the Extended Memory Board. If you will follow and study the circuit foils that lead away from the solder pin pads of IC-504, you will note unused solder tabs on most of the traces. We will connect most of the wires from the Extended Memory Board to these unused tabs instead of soldering directly to IC-504's pin pads. Actually, these "unused" solder pads are "plated thru" holes that continue the trace on the opposite side of the PCB. For our purposes, they are called (and will serve as) "unused solder pads."

For example, locate Pin 2 where IC-504 used to be. (Pin 1 is clearly marked on the PCB.) Note, with the front of the scanner facing you and viewed from above, that the trace from Pin 2 leads off to the right for maybe 1/4" and then bends 90° towards the front of the scanner for another 1/2" or so where it terminates (ends) at an unused solder pad. Using the wiring diagram in Figure 4-16-1, you will measure and fit the wire from Pin 4 of the Extended Memory Board to this unused solder pad of IC-504, Pin 2 just described above.

The extra solder pads for IC-504, Pins 3 through 8 are off to the left and slightly upper-left of IC-504 near and just to the rear of "D-515" which is marked on the board. Extra pads for IC-504's Pins 9, 10, and 11 are just to the right of IC-504. The wire from the Extended Memory Board's Pin 14 will go to the same printed circuit ground spot as the wire soldered in Step 23 above. Extra solder pads for IC-504's Pins

13, 14, and 15 are located within the the marked rectangular area of where IC-504 used to be located. Extra solder pads for IC-504's Pins 16 through 23 are located in a diagonal line just forward of where IC-504 was formerly located. The wire from the Extended Memory Chip's Pin 28 will solder directly to the IC-504 Pin 24 pad since there isn't an extra pad for this pin.

Essentially, then, the only solder pads that have to used where IC-504 used to be situated are Pins 1 and 24, and these are larger than normal pads anyway, which offer no problem. Thus, you can see that all the wiring rrom the Extended Memory Board to the vicinity of IC-504 will be fairly well spaced out and not too difficult. If you need further clarification, refer to the wiring diagram in Figure 4-16-1 and perform the detailed instructions explained in Steps 30 to 53 which will tell you exactly where to solder the wires in the PRO-2004.

Special Note For PRO-2005 Owners: Judging from a close scrutiny of the Service Manual, there are larger, unused solder pads (plated thru holes) for most of the pins of IC-505. You'll have to search them out using the Service Manual and by actually looking at the circuit board itself. The actual pinout of IC-505 is identical to that described below for the PRO-2004. Therefore the electrical pinout points of installation for the wires from the Extended Memory Board are valid. The physical locations of any unused solder pads will be different. It's OK to solder the wires directly to IC-505's empty pin pads, if necessary. It's just a little more tricky and tedius that way because the pins are located so close together.

Now to Steps 30 to 53, showing the exact wiring information in the PRO-2004.

For the sake of brevity in these steps, "MCST&S" will mean "measure, clip, strip, tin, and solder," and "EMB" will mean "Extended Memory Board." When clipping each wire to fit, allow a little slack for bundling, routing, and neatness.

30. MCST&S the wire from the EMB's Pin 3 directly to IC-504 Pin 1.

31. MCST&S the wire from the EMB Pin 4 to IC-504 Pin 2. Note: follow the trace from IC-504 Pin 2 off to the right and forward a little to the larger, unused solder pad. Solder the wire to this pad rather than directly to Pin 2.

32. MCST&S the wire from the EMB Pin 5 to an unused pad on the circuit trace of IC-504 Pin 3. This spot will be the anode of D-515 which is marked on the board. D-515 works in conjunction with D-514 to program the CPU for different speeds. If you should remove D-515 (don't do it), the scan and search speeds would slow down.

33. MCST&S the wire from the EMB Pin 6 to an unused pad on the circuit trace of IC-504 Pin 4. This spot is at what would be the anode of D-514 if you installed it for the scan-rate speedup described in MOD-2 of this book.

34. MCST&S the wire from the EMB Pin 7 to an unused pad on the circuit trace of IC-504 Pin 5. This spot is the anode for D-513, which is not marked on the board. D-513 is the cellular car phone elimination diode discussed in MOD-1 of this book.

35. MCST&S the wire from the EMB Pin 8 to an unused pad on the circuit trace of of IC-504 Pin 6. This spot is the anode of D-512 which is not installed but is marked on the board. As incidential trivia, should D-512 be installed, scanner operation will be deleted in the 30 to 54 MHz band, as required by Europe and Australia.

36. MCST&S the wire from the EMB Pin 9 to an unused pad on the circuit trace of IC-504 Pin 7. This is the spot for the anode of D-511 which is neither used nor marked on the board. Even when a diode is installed at D-511, it doesn't appear to accomplish anything.

37. MCST&S the wire from the EMB Pin 10 to an unused pad on the circuit trace of IC-504 Pin 8. This is the spot for the anode of D-510, which you may have already added to give you 100 more memory channels (see MOD-15).

38. MCST&S the wire from the EMB Pin 11 to an unused pad on the circuit trace of IC-504 Pin 9. This spot is located off to the right of IC-504, near Pin 3.

39. MCST&S the wire from the EMB Pin 12 to an unused pad on the circuit trace of IC-504 Pin 10. This spot is located off to the right of IC-504 near Pins 4 and 5.

40. MCST&S the wire from the EMB Pin 13 to an unused pad on the circuit trace of IC-504 Pin 11. This spot is located off to the right of IC-504, near Pin 6.

41. MCST&S the wire from the EMB Pin 14 to the same printed circuit ground spot as the wire soldered in Step 23.

42. MCST&S the wire from the EMB Pin 15 to an unused pad on the circuit trace of IC-504 Pin 13. This spot is located within the upper-left rectangle of where IC-504 had been previously positioned.

43. MCST&S the wire from the EMB Pin 16 to an unused pad on the circuit trace of IC-504 Pin 14. This spot is located within the upper-left rectangle of where IC-504 had been previously positioned.

44. MCST&S the wire from the EMB Pin 17 to an unused pad on the circuit trace of IC-504 Pin 15. This spot is located within the upper-left rectangle of where IC-504 used to reside.

45. MCST&S the wire from the EMB Pin 18 to an unused pad on the circuit trace of IC-504 Pin 16. This spot is located just forward of IC-504's rectangle and between the CPU chip and the row of diodes and resistors off to the left area of the CPU board.

46. MCST&S the wire from the EMB Pin 19 to an unused pad on the circuit trace of IC-504 Pin 17. This spot is located just forward of IC-504's rectangle and between the CPU chip and the row of diodes and resistors off to the left area of the CPU board.

47. MCST&S the wire from the EMB Pin 20 to an unused pad on the circuit trace of IC-504 Pin 18. This spot is located just forward of IC-504's rectangle and between the CPU chip and the row of diodes and resistors off to the left area of the CPU board.

48. MCST&S the wire from the EMB Pin 21 to an unused pad on the circuit trace of IC-504 Pin 19. This spot is located just forward of IC-504's rectangle and between the CPU chip and the row of diodes and resistors off to the left area of the CPU board.

49. MCST&S the wire from the EMB Pin 22 to an unused pad on the circuit trace of IC-504 Pin 20. This spot is located just forward of IC-504's rectangle and between the CPU chip and the row of diodes and resistors off to the left area of the CPU board.

50. MCST&S the wire from the EMB Pin 24 to an unused pad on the circuit trace of IC-504 Pin 22. This spot is located just forward of IC-504's rectangle and between the CPU chip and the row of diodes and resistors off to the left area of the CPU board.

51. MCST&S the wire from the EMB Pin 25 to an unused pad on the circuit trace of IC-504 Pin 23. This spot is located just forward of IC-504's rectangle and between the CPU chip and the row of diodes and resistors off to the left area of the CPU board.

52. MCST&S the wire from the EMB Pin 27 to an unused pad on the circuit trace of IC-504 Pin 21. This spot is located just forward of IC-504's rectangle and between the CPU chip and the row of diodes and resistors off to the left area of the CPU board.

53. MCST&S the wire from the EMB Pin 28 directly to the original solder pad of IC-504 Pin 24.

This completes the wiring in the vicinity of IC-504. At this time, check, double check, and **triple** check all wiring performed in Steps 30 to 53. Test each wire and solder joint for a firm, secure connection. Use a bright light and a magnifying glass to inspect each connection in minute and nit-picky detail. A careless mistake at this point could be frustrating, inconvenient, time consuming, financially expensive-- and, even worse, could cost you time spent using your scanner. So, take the time to verify your work.

Final Installation:

54. Refresh your memory of Step 2 above and then retrieve the four loose wires left aside in Steps 16 and 23 and solder them to the Extended Memory Board as follows:

    A. The loose wire from J-2 socket 4 (farthest right, facing the scanner) to the Extended Memory Board Pin 23, at the jumper stub.
    B. The loose wire from J-2 socket 3 to Extended Memory Board Pin 2 at the jumper stub.
    C. The loose wire from J-2 socket 2 to Extended Memory Board Pin 26 at the jumper stub.
    D. The loose wire J-2 socket 1 (farthest left) to Extended Memory Board Pin 1 at the jumper stub.

55. Insert a 4-segment DIP switch (S-1) with all segments "off" into J-2, the 8-pin IC socket.

Installation of The Extended Memory Chip:

56. Now **very carefully** remove the Extended Memory Chip from its protective packaging. Lift the chip by its ends from the carbon foam. Avoid touching the pins of the chip. Temporarily position the chip over its socket on the Extended Memory Board to see if the pins line up with the socket holes. Chances are, they will not fit perfectly at this point because the two rows of pins will be spread too far apart.

In no case should you attempt to force the chip into the socket! Instead, determine why the pins don't make perfect alignment with the holes and make corrective adjustments. Most of the time, the two rows of pins on the chip are too far apart-- and here is the remedy for that:

Grasp the Extended Memory Chip by its ends and lay it on a side so that all the pins in one row are flat against a hard surface. Torque the chip in such a direction as to cause the pins to all bend inward by the same amount at the same time. Turn the chip over to its other side and repeat this procedure. When properly aligned, the two rows of pins should be perfectly parallel with each other and perpendicular to the flat sides of the chip.

Again position the memory chip over the socket and if alignment of the pins is nearly perfect with the socket holes, then gently push and work the chip down into the socket. Be watchful and careful that the pins don't buckle and crimp as the chip goes into the socket. At this time, be very sure that the #1 pin is properly oriented. The chip can fit the socket in either of two directions, but only one way is correct. There is a "dot" on one corner of the chip to indicate the #1 pin. Since you wired the chip socket, you will know which pin of the socket is #1. Figure 4-16-1 shows the pin numbering configuration.

The work is done now! We're about ready to energize the scanner for some performance tests. Use this opportunity to review your work one more time before proceeding.

Table 4-16-4

INITIAL PROGRAMMING OF 400-CHANNEL BLOC IDENTIFIERS

	DIP SW #1	DIP SW #2	DIP SW #3	DIP SW #4	PROGRAM CH-1	=	400-Ch MEMORY BLOC #	BINARY EQUIV
Home	OFF	OFF	OFF	OFF	1000.000	=	00 Home	0000
	OFF	OFF	OFF	ON	1001.000	=	01	0001
	OFF	OFF	ON	OFF	1002.000	=	02	0010
	OFF	OFF	ON	ON	1003.000	=	03	0011
	OFF	ON	OFF	OFF	1004.000	=	04	0100
	OFF	ON	OFF	ON	1005.000	=	05	0101
	OFF	ON	ON	OFF	1006.000	=	06	0110
	OFF	ON	ON	ON	1007.000	=	07	0111
	ON	OFF	OFF	OFF	1008.000	=	08	1000
	ON	OFF	OFF	ON	1009.000	=	09	1001
	ON	OFF	ON	OFF	1010.000	=	10	1010
	ON	OFF	ON	ON	1011.000	=	11	1011
	ON	ON	OFF	OFF	1012.000	=	12	1100
	ON	ON	OFF	ON	1013.000	=	13	1101
	ON	ON	ON	OFF	1014.000	=	14	1110
	ON	ON	ON	ON	1015.000	=	15	1111

Final Tests and Checkout:

57. Reinstall the Memory Retention Battery in the compartment on the rear of the scanner. If you haven't already done so, set all four segments of the DIP switch to the "off" positions.

58. Plug the scanner into appropriate AC or DC power.

59. Turn the scanner on and observe all indications for "normalcy." Quickly shut the power off if there are any peculiar symptoms, otherwise punch up Manual : 1 : Manual. Channel 1 ("manual" mode) should now be in the display. Operate the "scan" and other keys to ensure proper operation of all keyboard functions. If everything appears normal, move on to Step 60.

Note: Some or all channels in each of the 16 blocs of 400 channels might appear to be programmed with frequencies, including some really bizarre ones along the lines of "5345.6873" and similar. This is caused by "noise" that entered the chip before and during installation. Use the "Restart" switch on the rear chassis to clear all 16 blocs before reprogramming. The "Restart" switch will clear only one 400-channel bloc at a time, so you'll have to do it 16 times, with the DIP switch in each of its 16 positions. It's a good idea to clear the memory of all of this clutter so you won't become confused later when you find it impossible to keep track of 6,400 channels. This subject and some ideas for "Scanner Channel and Frequency Management" are discussed in the chapter of this book on "Tips, Hints & Kinks."

60. Press Program : 1 : Program 1000.000 : Enter. Frequency 1000.000 MHz should now be displayed on Channel 1.

61. Set the #4 segment (farthest right) of the DIP switch to the "on" position and press Program : 1 : Program 1001.000 : Enter. Frequency 1001.000 MHz should now be displayed on Channel 1.

62. Turn the #4 segment of the DIP switch to the "off" position and press Manual : 1 : Manual. Frequency 1000.000 MHz should again be displayed. Reset the #4 segment of the DIP switch back to the "on" position and again press Manual : 1 : Manual. Frequency 1001.000 MHz should return to the display.

Note: This demonstrates two different blocs of 400 channels each. With all four DIP switches in the "off" position, this should be loosely referred to as the "home" or stock 400-channel memory bloc. When the DIP switch is removed from the socket, the "home" memory bloc will always be active. We'll call this "home" memory bloc the "00" memory bloc. When DIP switch #4 is "on" and 1, 2, and 3 are "off," that will be the "01" memory bloc, and so forth.

63. If all is well so far, program Channel 1 of each of the 16 400-channel memory blocs as indicated in Table 4-16-4. This Table shows a special coded frequency you can program into Channel 1 in each of the sixteen 400-channel memory blocs. The remaining 399 channels in each memory bloc can be programmed with anything you want that the scanner will accept. The coded frequency in each Channel 1 will tell you at a quick glance which memory bloc the scanner is working from.

64. Note that because of the wires from the Extended Memory Board going into the CPU compartment, you'll not be able to replace the metal cover to the CPU compartment. Frankly, this isn't mandatory, but some people like to see all shields and covers put back where they were when the set came off the production line. Bundle the wires loosely together and measure the CPU cover so that you can determine where to cut away some metal. Use a "nibbling tool" to chew away enough of the metal so that the wire bundle can pass into the CPU compartment without chafing or abrading. Replace the metal cover. See the photos.

65. Disconnect scanner from its power source, replace the case, and restore it to its operating position where you can again plug it into its power source.

Wrapup and Final Discussion. Until you become accustomed to the binary coding of the 4-segment DIP switch, the programming of Channel 1 in each of the 16 memory blocs (per Table 4-16-4) will enable you to tell which bloc you are in at any given time. Just "Manual" to Channel 1. Note that any time you unplug the DIP switch from J-2, the scanner will automatically revert to memory bloc "00," which is the "home" bloc. This is the same as if all four switches are "off."

By the way, you can unplug the DIP switch to change its settings and then plug it back in, especially if it's difficult to see or reach in the scanner's regular operating position. Or, perhaps better still, you could even purchase several 4-segment DIP switches and preset each one to a desired memory block. Use a permanent marking pen to annotate which bloc number the switch is configured for on the front side of the DIP switch for identification at a glance. Actually, once you get acclimated to binary coding, resetting the DIP switch should become rather routine.

At this time, if no problems have reared their ugly heads, you will have 16 memory blocs of 400-channels each. Each memory bloc will also have 10 search bands and 10 temporary "monitor" (scratch pad) channels. It's almost like you went out and purchased 16 400-channel scanners-- except that you can't run them all at once. Still, not a bad trade-off for having 6,400 channels of scanning power. You are now free to program these 6,400 channels in any manner you wish so long as the frequencies lie within the capabilities of the CPU. You will now want to work up a master plan or intelligent scheme for keeping track of what you'll be programming into your scanner. You could devote one entire bloc of channels to police, another to federal, another to VHF aero, another to UHF aero, another to--- well, anything of your choice!

One of the strong points in favor of this modification is that you can customize each of the 6,400 channels for mode (AM, NFM, or WFM), and for delay, since this data is stored in the Extended Memory Chip along with the channel information.

A minor limitation here is that the "priority" and "speed" functions are stored in the CPU chip, which cannot be modified. Therefore, when you set a PRI Channel or change speed for any given 400-channel bloc, that same channel will be the PRI channel for the other 15 400-channel memory blocs-- and the speed will be the same for all blocs. Naturally, you can have 16 different PRI frequencies, but they must all be stored in the same channel number. To keep things from getting confusing, you might just allocate Channel 2 in each of the 400-channel memory blocs for PRI status. This one limitation is the only drawback to the modification that I have encountered, and it's not actually a real drawback, it's just a lack of flexibility.

MOD-17 68 to 88 MHz & 806 to 960 MHz for The Realistic PRO-2021-- Well, Maybe!

Well, this is one of those deals where "they all said it couldn't be done" but some people insist that they've done it and it works. Others report that they tried the same thing and it dodn't work. The claim is that the Realistic PRO-2021 scanner is capable of operation on 66 to 88 MHz and 806 to 960 MHz when certain diodes are clipped. You won't have to go to a lot of trouble to find out for yourself, but first let me clue you in on some of the odd things about claims that this could actually work.

When it comes to liberating repressed bands or frequencies from a modern scanner, there are two distinctly separate areas of the scanner that have to be inherently capable (by design) of the coverage in question. One is the microprocessor (CPU) and the question of whether or not it had been factory-programmed so that there is a capability within the CPU for such capability. If the CPU wasn't <u>originally programmed</u> by the factory programmed for such coverage, then let's put out the lights and go to sleep, because you can forget the whole idea of a modification. You're out of luck!

On the other hand, sometimes the CPU does contain some weird programming that doesn't seem to be supported out in front by the keyboard. Cases in point: Realistic PRO-34, PRO-2004, and PRO-2005. The CPU's are programmed to receive the cellular bands. For reasons of their own, manufacturers sometimes prevent keyboard access to certain coverage by a "secret" diode wired in (or omitted) across two relevant terminals of the CPU. If you can find and remove or add that diode, as appropriate, you <u>might</u> just be in luck. Might??

Yup, you're still not quite home free. The second critical area of the scanner that has to be capable (by design) of allowing expanded frequency modifications is the RF circuit which actually consists of a couple of different areas. For purposes of this discussion, however, I'm going to treat them as if they were one.

All scanners have RF preselector, frequency mixer and oscillator circuits that are designed to be operational on specific bands of frequencies. For example, the

PRO-2004/2005 supports seven such bands, designed as follows: 25 to 40 MHz, 40 to 68 MHz, 68 to 108 MHz, 108 to 174 MHz, 174 to 280 MHz, 280 to 520 MHz, and 760 to 1300 MHz.

Observe that the band 520 to 760 MHz does not exist, which is why reception of this band (UHF TV Channels 22 through 62) is probably impossible, even if we could find its secret programming in the CPU. The PRO-2004/2005 simply doesn't possess the RF circuitry to accept signals between 520 and 760 MHz. And, so too it is with other scanners in regard to various frequency bands, even though the programming in the CPU <u>might</u> exist.

For example, I was able to "liberate" 800 MHz to the display in the older PRO-2002 base and later PRO-32 handheld scanners, but in neither case was I able to actually receive any 800 MHz band signals. To be sure, after I clipped or added certain diodes, I was able to use the keyboard to seemingly program, scan, and search 800 MHz, but never was I able to actually receive anything. A cursory look at the schematic diagrams revealed why not-- the RF circuits for 800 MHz didn't exist. I did manage to uncover a little surprise in the PRO-2002 when I found the CPU "secret" programming for 380 to 410 MHz. As it turned out, the factory designed RF circuits for 410 to 512 MHz were wide enough to lower the coverage of 380 to 410 MHz, and I was thereafter able to utilize the scanner here, which was primarily of value in the 406 to 410 MHz area, which is used by federal government agencies.

So, here's the deal for the PRO-2021. The CPU is apparently factory programmed for 800 MHz coverage, maybe because they used the same CPU chip used in other Realistic scanners that were designed to work on 800 MHz. Use of the same parts in several different models is sometimes a wise economy move that manufacturers employ. In any event, from the PRO-2021's I have seen, the model does not appear to have any 800 MHz circuitry and so it <u>should be impossible</u> to receive anything even if you can get 800 MHz frequencies programmed into the LCD display. You'll just have to find out for yourself because some people insist they can receive signals while others swear they can't. Maybe there are several "versions" of the PRO-2021-- I don't know. Here's the modification that has been employed:

806 to 960 MHz Restoration for PRO-2021 (Questionable)

1. Disconnect scanner from its power source. Remove it from its case and find your way to the CPU circuit. Locate and clip one leg of D-44.

2. Program in the limits of 851.000 to 866.000 MHz, for example, into a "search" band. Activate the "search" and if you hear the various two-way communications services that operate there, the mod worked for you. If you don't receive anything, well, no harm done.

68 to 88 MHz Restoration for PRO-2021 (Maybe)

Some folks also assert that 68 to 88 MHz can be reclaimed from the PRO-2021. The same discussion from above applies here, maybe it can and maybe it can't. My guess is that it could happen since the PRO-2021 has RF and digital capability of 30 to 54 MHz which isn't all that distant from 66 to 88 MHz. Beware, however, that when you clip that diode, you might add 68 to 88 MHz (which is of minor monitoring interest, at best) at the cost of losing 30 to 54 MHz or something else you might really want. Fully test your scanner after the mod to make sure:

1. Disconnect scanner from its power source. Remove it from its case and find your way to the CPU circuit. Locate and clip one leg of D-45.

2. Program the limits of 68.000 to 88.000 into a "search" band. Activate the "search" and let things run for a while to see if you can pick up anything. The major portion of this band is occupied by TV Channels 4, 5, and 6, but between 72 and 76 MHz, if your mod worked, you might pick up some telemetry and radio control tone signals, also perhaps voice and non-voice radio paging relay transmitters. The audio

carriers for the TV channels are: Channel 4= 71.750 MHz; Channel 5= 81.750 MHz; Channel 6= 87.850 MHz. If your scanner doesn't pick any of this up, no harm done-- unless the removal of D-45 blocked out something else.

MOD-18 **Restoring The Removed Cellular Bands in The Realistic PRO-34 Handheld Scanner**

This modification is a bit hairy because of the major mechanical and electrical disassembly necessary to gain access to the teeny-weeny, little, diode that has to be clipped. If you are patient, methodical and careful, and you don't mind learning how an elephant feels in a telephone booth, then it can be done with nary a hitch.

But Read This First: Rumors have circulated about a recent manufacturing change where a metal board-shield inside the PRO-34 is now spot-welded to the main frame instead of just being held down with screws. They say that the welds have to be popped loose now in order to access the CPU and Logic board. I haven't seen this, but the possibility is worth passing along.

To do this mod, you'll need a soldering iron, desoldering tools, small wire cutters, small adjustable wrench or various miniature open-end wrenches, small-tipped phillips screwdriver, and something to pry with. Ready? No, you will first want to review the all-important Service Manual pictorials and exploded views to get familiar with your mission before you embark.

Caution: Chances are that you won't lose what is in the scanner's memory if the work is completed within an hour or so. Memory is retained by a large internal capacitor when the batteries are removed. As a backup, you might wish to write down a listing of the frequencies you have programmed in the PRO-34 just in case it comes down with a sudden attack of amnesia.

1. Remove the antenna, battery pack, and the "volume" and "squelch" knobs. The knobs might have to be removed with the aid of pliers.

2. Remove the 4 small phillips screws from the rear cover.

3. Starting at the battery-pack end of the scanner, pry the back cover loose. Do

Taking apart the PRO-34 for restoring its cellular operation takes some patience, deft fingers, good eyesight.

Location of D-11 in the PRO-34, the cellular lockout diode. It's tiny.

not attempt to pry at the top (it's hinged there). Swing the back cover up in an arc from the bottom end while pushing toward the top of the radio. The "hinge" at the top will work loose and the rear cover can be worked off over the antenna connector and the "volume" and "squelch" shafts. Lay the rear cover aside.

4. Loosen and remove the 4 brass hex shaft/spacer screws from the exposed printed circuit board.

5. Note the "volume" control and the two bare wires going from it to the circuit board. Desolder those bare wires <u>at the circuit board</u> and push them up out of the way against the body of the volume control.

6. Locate the antenna connector where it solders to the circuit board in two places. Desolder the antenna connector ground-lug from the circuit board and bend it back out of the way. Use a solder sucker to remove the solder "blob" at the center conductor of the antenna connector at the circuit board.

7. Looking at the rear of the radio (exposed circuit board) with the the antenna pointed up, locate the small, bare ground-wire on the right side of the radio, about halfway down from the top. Actually, it is at the lower-right corner of the exposed circuit board. This wire connects the metal can of a small transformer to a metal shield cover below. Desolder that wire from the metal shield cover.

8. Loosen and remove the two hex nuts and washers from the shafts of the "volume" and "squelch" controls.

9. Carefully, but firmly, pry the exposed circuit board up. It is held in place by a connector hidden beneath it. Use a small flat object to pry with, using the sides of the plastic case as a fulcrum. Pry only near the top end of the board on either side where two brass hex shaft/spacer screws were removed. If you look on the board between where the two brass screws were, you'll see a line of "pins" which are the back side of the connector that has to be pried up. You'll feel a little "give" as the board works up and loose from the connector. As the circuit board works up, you'll be able to see the connector hidden under the board. Continue working the exposed circuit board up, straight away from the body of the radio until it comes loose from the connector beneath. Gently work the board so that the "volume" and "squelch" controls slip out of their mounting holes as the board is lifted away. Remove the small connector with the red and white wires from the inner circuit board that goes to the now-removed upper circuit board. The entire upper board should now be free and placed aside in a safe place.

10. Note the shield cover now fully exposed within the body of the radio. It must be removed. This is the shield can that some rumors claim is now being spotwelded on

later models instead of being screwed-down. I can't vouch for that, so if yours is spotwelded, you're on your own. Apparently the spot welds can be popped loose-- or so they claim, at least.

If your shield isn't welded, then it will be held in place by three small phillips screws, one at the lower edge of the shield next to the battery compartment, and the others, on the other side of the shield cover near the upper end (of the shield cover). Remove these three tiny screws and lift the shield cover out of the way to expose the CPU and Logic board beneath. The shield will not come completely free because of some wiring and the PWR and CHG jacks. Just move the cover over and lay it next to the radio.

11. Congratulations! We're halfway done and getting down to the nitty gritty. Observe to the left side of the CPU and Logic board (the side opposite the PWR and CHG jacks), and locate diodes D-10 and D-11. These are located on the extreme left edge of the logic board, and just to the left of IC-2. Clip the exposed leg of D-11 and spread the cut ends slightly apart so they can't touch. Don't fool around with D-10 or else you'll lose 900 Mhz! Just clip (only) D-11 and you'll be done and ready to button things up.

12. Carefully replace the metal shield cover so that it aligns with the three screw holes. Replace and tighten the three screws.

Note: It would help a lot to magnetize the shaft of your phillips screwdriver for this operation. Just wipe or stroke a magnet along the shaft about 6 to 10 times in one direction only. The phillips screwdriver will now hold the tiny screws on its point so you can more easily align them with the holes in the shield cover. Tighten the three screws.

13. Gently reposition the top circuit board over the metal shield cover and replace the small connector (CN-1), the one with the red/white wires. Work the "volume" and and "squelch" controls back through the holes in the top frame, and then position the board so that its male connector pins on the bottom side align perfectly to match with the female connector on the CPU and Logic board. When the pins are lined up, work the upper circuit board down until it seats. It would help to press with your thumbs directly over the connector pins.

14. Replace and tighten the 4 brass hex shaft/spacer screws.

15. Replace and tighten the hex nuts and flat washers over the "volume" and "squelch" controls.

16. Bend the two wires from the "volume" control back down to their spots on the circuit board and resolder them to the board.

17. Resolder the antenna connector's ground lug back to its spot on the circuit board.

18. Resolder the center conductor of the antenna connector back to its spot on the circuit board.

Note: There is a gap between the antenna connector center-pin and the solder spot on the board that might be very difficult to fill with solder without excessive heat and contact time. So, position a small bare wire about 3/16" long across the gap and then solder. There had been just such a tiny wire present there when you desoldered that spot, but it probably got lost in the shuffle and you may not have noticed it, anyway.

19. Resolder the interconnecting ground wire removed in Step 7.

20. Replace the rear cover, hinging it at the top first, before swinging it down into place. It will snap back into place with slight pressure applied along its length on both sides.

21. Replace the battery pack and the battery compartment cover.

22. Test the unit by programming the limits 880.650 and 889.980 MHz into the "search" bank. The set should be operational on these frequencies-- although it's a violation of the ECPA to actually listen to any of the cellular calls; see the chapter on "Scanning & The Law" and also the one on "Cellular Mobile Telephone - Explained." Also see the preliminary discussion given with the modification for restoring these frequencies in the PRO-2004/2005.

MOD-19   3,200 Programmable Channels for The Realistic PRO-34 Handheld Scanner

At first, I wasn't going to include this modification because it's tricky and because there is so much that could go wrong for a novice hacker. But, in retrospect, my position is not to decide what someone can and cannot do. You have to decide for yourself and then reap the rewards or pay the consequences. Still, it is not recommended for those who don't have lots of experience and know-how relating to circuits and how to work with them. Words to the wise-- I hope!

After loading my PRO-2004 with 6,400 channels, I was so elated and proud of my accomplishment that I was practically glowing for days. Then, after a few days, I began wondering what new worlds there were to conquer. That's when I reviewed the Service Manual for the PRO-34 and discovered that this fine handheld scanner used the identical SRAM memory chip (TC5517CF-20) used in the PRO-2004/2005. How's them apples? My rapidly-approaching depression immediately reversed gears as it occured to me that maybe there was a way of adding lots more channel space to the PRO-34 by the same route used in the PRO-2004.

The next day was filled with planning and preparation, catalogs, spec sheets, schematics, and calculations. No way was this going to be easy. Looking back now, it is possible that an enterprising hacker could make a simpler and more aesthetic project of it than I did. Nevertheless, the job was a complete success, and the results (although not on a par with the PRO-2004/2005) were worth the trouble. But then, I'm a rather dedicated and totally committed communications buff, so 3,200 channels in my handheld scanner is a definite asset!

An external view of the PRO-34 set up for 3,200 channel operation. The external cable bundle (and the Memory Chop) can probably be wired and routed internally if you are willing to use external power and forget the battery pack.

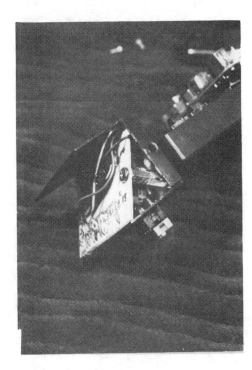

Things are a bit cramped inside the metal box, but perfectly workable. The 24-wire bundle is routed out the opposite side of the box (not shown).

I noted that the results were not on a par with the PRO-2004/2005 mod for 6,400 channels. I said this with good reason, and it's not because you're going to get only half the number of channels. See, the PRO-34's CPU chip (like most CPU chips) has a small memory bank of its own called **R**andom **A**ccess **M**emory (RAM). It's not much, but certain keyboard programmable information is stored in the CPU's RAM. In the case of the PRO-2004/2005, "priority" and "speed" happen to be stored in the CPU's RAM. Fortunately, "search limits," "monitor channels," "modes" (AM, NFM, and WFM), "lockout," and "delay" are all stored in the SRAM chip. That's wonderful, because it permits custom programming of all these functions from one bloc to another, each one totally unique from the others.

Not so in the PRO-34 which relies on the SRAM only for programmed "scan" frequencies. All the good keyboard stuff is locked up in the CPU's RAM. So what this means is if you could add a 256K SRAM to the PRO-34, all you're going to get out of it are a piddling 3,200 channels. (Piddling???) Now let's say you select the first bloc of 200 channels and set "delay" for Channel 3 and "lockout" for Channel 7. Well, Channel 3 for the remaining 15 blocs of 200-channels will also have the "delay" set, and all 16 Channel 7's will be locked out! Mind you, we're not losing anything here; there is the gain of 3,000 more programmable channels-- and that's rather awesome in itself. Just accept the fact that when you custom set a given channel in the PRO-34, then all 16 blocs of that same channel number will be set up for "delay" or "lockout." Furthermore, you'll still get only 10 "monitor" channels and one "limit search" band, since these, too, are stored in the PRO-34's CPU RAM. If that's OK with you, then let's get on with the hairy work facing us.

Preliminary Instructions. I'm not going to give the detailed instructions here that I did for the PRO-2004/2005. You can read these Steps of Procedure, most of which would be repetitive here anyway. Since the stock memory SRAM chips are identical among the PRO-2004/2005 and the PRO-34, the wiring connections will be identical. The required 256K chip and the Extended Memory Board are also identical. In fact, there are only a few differences for between the PRO-34 memory modification and the one for the PRO-2004/2005. These differences and keynotes only will be pointed out. Refer to the photographs for this modification, and that should be very useful-- otherwise apply the discussion and instructions in MOD-16 for the PRO-2004/2005 to the PRO-34.

PRO-34 3,200-Channel Discussion/Instructions:

1. Read all the information for MOD-16 because much is explained there that applies here, too.

2. Use Figure 4-16-1 (wiring diagram) and the detailed instructions for fabricating the Extended Memory Board when you work on the PRO-34.

   A. In the PRO-34, the circuit symbol of the TC5517CF-20 SRAM memory is IC-2. In the PRO-2004 it was called IC-504, so substitute "IC-2" in Figure 4-16-1 (wiring diagram) where ever reference is made to IC-504. Of course, the pin numbers do not change.

3. Use Tables 4-16-2 (parts list) and 4-16-3 (tools and materials list).

   A. Fabricate the Extended Memory Board in accordance with instructions given in MOD-16 for the PRO-2004.

   B. For the PRO-34, the Extended Memory Board (EMB) will have to be mounted outside the radio. That, or else install it in the battery compartment and then use external batteries or an outside power source. If you decide to install the EMB outside the radio, you'll need a small metal box in which to enclose the EMB. Add that box to your parts list, together with some small machine nuts and bolts (Radio Shack #64-3011 and #64-3018).

   C. The metal box chan be mounted on the bottom-end of the radio, provided you offset it just a little toward the front side of the radio so the battery compartment lid can still slide off. (The radio can still be stood up on end this way). The metal box can be mounted with two #4-40, 1/4" long machine screws and nuts. Drill the two screw holes through the bottom-end of the radio into the battery compartment. Position the two holes so that the battery pack does not hit the screw heads when it is inserted. The bottom edges of the battery pack are rounded, so this won't be a major problem. The rounded edges will clear the screw heads if the screw holes are positioned deep enough into the bottom-end of the radio.

   Two mating holes will have to be drilled into the metal box such that the box will be offset toward the front of the radio, to give clearance for the battery lid to be slid on and off the back.

   Mount the EMB inside the metal box using spacers or shims and a machine screw and nut.

   The 4-segment DIP switch should mount on the <u>front</u> side of the metal box.

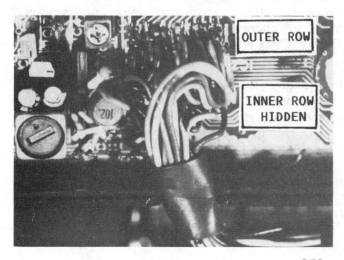

All wiring of the Extended Memory Board in the PRO-34 is done at the pins of IC-2. Wire up the inner row of pins before the outer row.

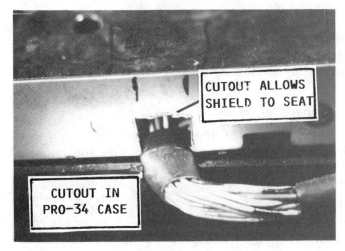

Access hole in the PRO-34 for the Extended Memory Board wire bundle. A narrow cut is made in both the plastic outer case and the inner metal shield cover as shown. The metal shield cover is shown partially raised to expose the cutaway section.

Review the photos for ideas on how to route the wiring bundle out of the metal box and into the side of the PRO-34.

4. In the PRO-34, because of micro-mini printed circuitry, there is no alternative to soldering the 24 wires from the EMB directly to the original solder pads where IC-2 will be removed. In the PRO-2004, we soldered the wires to unused solder pads, but we don't have that space luxury in the PRO-34. This does not pose a severe problem, but there is no other alternative. IC-2 has to be removed and the 24 wires from the EMB must be appropriately soldered to IC-2's 24 pin pads.

5. Use the disassembly procedures given in MOD-18.

6. Since the wiring bundle from the EMB must enter the radio on the left side as you face the rear of the PRO-34, a hole must be made on the side of the radio. Not only must a hole be cut in the side of the plastic case, but a matching hole must be cut in the internal metal shield cover (which you will encounter when the scanner is disassembled). Actually, this isn't difficult and it looks pretty good (see photos). I used a "nibbling tool" to cut matching rectangular slots in the plastic side of the scanner and in the internal metal shield cover for the wiring bundle.

7. Remove IC-2 from the CPU and Logic board using standard desoldering techniques and relevant instructions given for the PRO-2004 memory modification.

8. Install the metal box on the bottom-end of the PRO-34. Install the EMB inside the metal box. Install the 4-segment DIP switch on the front of the metal box. You can use an 8-pin DIP socket just like that specified in MOD-16 for the PRO-2004/2005. Wire the 4-segment DIP switch (or its socket) to the EMB board per Figure 4-16-1. Route the wiring bundle out of the metal box through a hole or slot appropriately prepared. Slip on at least five half-inch lengths of heat shrink tubing over the wiring bundle. Use a rubber grommet as an exit for the wire bundle to pass through the side of the metal box.

9. Route the wiring bundle into the side of the PRO-34 through a rubber grommet placed in the slots or holes cut for that purpose and pre-dress (measure, clip, strip, and tin) the wire ends for soldering to the pads where IC-2 was removed.

The wiring bundle should be pulled fairly tight and forced into a sharp 90° bend to enter the side of the radio.

You must solder the appropriate wires to IC-2's solder pads, pins 13-24 first-- before soldering wires to pins 1-14, otherwise you won't be able to access pins 13-24 because they'll be covered up.

Solder the 24 wires from the EMB in accordance with Figure 4-16-1 to the now vacant solder pads where IC-2 had been located, pins 13-24 first, then pins 1-14. Use a strong magnifying lens and a bright light to closely inspect all 24 pins where they are soldered. Make certain there are no loose connections or short circuits between pins!

10. Reassemble the radio using the steps given in MOD-18.

11. Refer to Table 4-16-4 for initial programming of the bloc identifiers, but use 900.0, 901.0, 902.0, etc. since the PRO-34 will not accept 1000 MHz. Use the general test procedures given for the PRO-2004 to test the 3,200 channel memory of the PRO-34.

12. Now you've completed this modification. Enjoy your PRO-34 with a feeling of pride and accomplishment. How many other 3,200 channel handheld scanners are there? Not too many-- for all we know, only two, yours and mine! Congratulations!

MOD-20   **Restoring Cellular Frequencies in The Uniden/Bearcat BC-950XLT**

What follows will liberate the whole 806 to 956 MHz band, which includes cellular, business band, public safety, ham radio frequencies, and more. You'll need a phillips screwdriver, fine-point soldering iron, rosin-core solder, and a fine-point cutting tool. Cuticle scissors or small diagonal cutting pliers will work. But remember, if you mess it up, Uniden doesn't recommend modifications to their equipment, and your warranty is voided.

1. Make sure the scanner is disconnected from its power source. Remove the 4 screws holding the bottom cover.

2. Remove bottom cover carefully, protecting the speaker wires.

3. Find microprocessor chip (64-pin IC at front of board).

4. Note the indented dot that marks Pin 1; count along that row to Pin 20 and cut it loose <u>at the board</u> (not at the IC body). Bend Pin 20 up a little so it can be worked with.

5. Solder a bridge at the chip between the clipped Pin 20 and Pin 19.

6. Reconnect the power to the scanner. Turn it on and program in a frequency between 806 and 956 MHz to confirm operation. If you get no reception, especially between 851 and 866 MHz or 870 and 890 MHz, check for accidental solder bridges to other pins or an incomplete disconnection of Pin 20. And away you go!

Note: If all else fails, remove the "bridge" you soldered between Pins 19 and 20 in Step 5, and test the scanner per Step 6 again.

Final Note: The "search" step increment will probably be 12.5 kHz, which would conform exactly with FCC frequency allocations in this range-- except for the cellular bands where the channels are spaced at 30 kHz increments. Receiver selectivity should be sufficiently wide enough so that you won't realize the difference.

MOD-21   **Restoring Cellular Frequencies in The Uniden/Bearcat BC-200/205XLT**

Again, Uniden doesn't recommend or endorse modifications to the innards of their scanners. If you could care less, you will need a small wire cutter or clipper and a small phillips screwdriver.

1. Slide off battery pack.

2. Remove the two screws from the rear of the scanner and the two screws that hold the battery retaining spring at the base. Then remove the spring.

3. Carefully pry the bottom of the rear cover from the scanner and remove the cover.

4. Locate the two small screws at the base of the circuit board and remove them. Gently pull the front panel from the main frame at the base and separate them.

5. Locate the microprocessor chip (marked "Uniden UC-1147") and the 10K ohm (browm/black/orange) leadless resistor located just above the letters "den" of "Uni**den**" on the chip.

6. Snip the resistor body in half with the small wire cutters. Take pains to avoid disturbing anything else in the vicinity. If the left solder pad comes loose, peel it off the board. Brush or blow away any residue. The restoration is complete.

7. Reassemble: Insert top of the front panel into slot under the volume/squelch control panel. Carefully noting the alignment of the dual in-line connector at the bottom of the board with the mating socket, press the front panel firmly into place. Be sure that the holes at the bottom of the circuit board line up with the holes in the plastic standoffs below them.

8. Replace the rear cover by inserting the top of the cover into the slot under the volume/squelch control panel; press cover into place. Insert and tighten the screws.

9. Reposition the battery retaining spring (slotted side toward notched hole). Insert the two remaining screws and gently tighten them.

10. Slide the battery pack into place. Switch radio on to make certain the display comes on. If not, the battery may be discharged or the dual in-line connector was misaligned diring reassembly (Step 7). Assuming that the display comes on, press: <u>Manual : 845.0 : E</u> and within a second or two, 845.000 should appear in the display.

Note: It should be pointed out that you might not have to actually go through Steps 1 through 10 since you can probably just take a stock (unmodified) unit and try receiving on the deleted frequencies via images-- as described earlier in this book. Try adding 21.7 MHz to the deleted frequency, and then programming in the higher (image) frequency. For instance, if you programmed in 908.860 MHz, reception of cellular frequency 887.160 MHz would occur, and with passable quality. So, any frequencies programmed in that lie above the deleted band and below 915.70 should produce reception on the bands that were deleted. It's worth a try, anyway!

MOD-22 — Restoring Cellular Frequencies in The Uniden/Bearcat BC-760XLT

If you are not concerned that Uniden does not recommend or suggest internal modifications to their scanners, and will cut you adrift from their warranty protection program if you go ahead anyway-- then you can free up the repressed 806 to 956 MHz coverage in the Bearcat BC-760XLT scanner. You will need a phillips screwdriver, fine-point soldering iron, rosin-core solder, and a fine-point cutting tool such as cuticle scissors or small diagonal cutting pliers.

1. Disconnect scanner from its power source. Remove the 4 screws holding the bottom cover.

2. Carefully remove bottom cover, protecting the speaker wires.

3. Find the microprocessor chip (64-pin IC at front of board).

4. Note the indented dot that marks Pin 1; count along that row to Pin 20 and snip it loose <u>at the board</u> (<u>not</u> at the IC body). Bend Pin 20 up a little. Test the scanner now per Step 5 for desired operation.

Note: Some sources report that this completes the modification. If, however, 800 MHz reception isn't restored, complete Step 4A below and test the scanner per Step 5 again.

4A. Solder a bridge at the chip between the clipped Pin 20 and Pin 19.

5. Reconnect the power to the scanner. Turn it on and program 880.030 MHz to confirm operation in the 800 MHz band.

Note: If operation wasn't restored, check for accidental solder bridges to other pins or an incomplete disconnection of Pin 20. That should find any problems. The scanner will probably search at 12.5 kHz increments, which should prove suitable for reception throught this frequency range despite the fact that some stations may be centered slightly "off center" frequencies.

MOD-23   Automated Search & Store Function for The PRO-2004 and PRO-2005

This is a nifty and useful addition to the functions of a PRO-2004/2005, but I can't tell you how to build it. There are two reasons why I can't provide the full and complete hacker details:

A. I don't know how to build it myself.

B. Design and construction of the modification board and circuit is the rightful property of a company.

But, fret not. I can still tell you about this, and let you know where to contact the company that offers this circuit.

A clever engineer has developed a small circuit board (sold in fully constructed and tested form) that allows your PRO-2004/2005 to "limit search" or "direct search" a band of your choice and then automatically store detected frequencies into the "monitor" channels, one at a time. What this means is that you can program and start a search band and then go to bed or to work. As the search function detects and stops on an active frequency, that frequency is automatically stored in a "monitor" channel, and the "monitor" channel number advances by one-- then the search automatically resumes until the next active frequency is detected. This is a fully automatic process that needs no attention from the operator.

Later, when you return, the new frequencies can be transferred to the permanent memory channels in the usual manner. The search/store module helps you find new or rarely used frequencies that would otherwise remain undiscovered for lack of the patience or the time to search out manually.

This module comes ready to install. Installation is simple (complete instructions are provided with the module) with 8 wires to solder to easily accessable points. No holes to drill, no wires to snip, no circuit patterns to cut. The installation took me about

The Key Research module #SS-45 Search and Store Module installed. Tucks away neatly behind the keyboard. All directions are supplied with the unit.

ten minutes-- but only because I'm super-careful and extra-critical of what I do. Most people would probably do it in less time.

Operation of the module is all via the keyboard (at the touch of a key) and can be activated or stopped at any time. All other operations are normal and intact. So, it's really much as if it came as a factory feature of the scanner. The module (called the SS-45) carries a 30-day "no questions" return guarantee, and a 90-day warranty against defects in parts and workmanship. At press time, it was priced in the $25 ballpark, but you might wish to inquire of the manufacturer of the current price and availability of the SS-45 (if you enclose a self-addressed, stamped return envelope, I'm sure it will be appreciated). The manufacturer is:

Key Research Company, P.O. Box 5054, Cary, NC 27511

The manufacturer advises that a second (more powerful) version is in the planning stages. Called the PS-90, it will have two modes of operation-- that of the SS-45, and a 2nd model that will store active "search" mode frequencies in the scanner's main channel memory. You could select (via an internal DIP switch) anywhere from 10 to 255 of the main memory channels to be allocated for "search" storage. The expected retail price is in the $45 area.

Either unit would match up well with MOD-16 in this book, the 6,400 memory channel modification-- or you could just use it with a 300 or (modified) 400 channel PRO-2004, or stock PRO-2005. I have no commercial ties with this manufacturer-- I just thought you'd like to know about their useful and interesting module(s).

MOD-24 **What The Future Holds in Store for Your PRO-2004, PRO-2005, & Other Scanners, too!**

The one with the answers is you. You are the customer who must proclaim what you want in your scanner! Necessity (and more often than not, demand) is truly the mother of invention. When you state what you want, then scanner hackers are inspired to theorize, hypothesize, and synthesize in order to develop what you want. Then the results come out in the hobby media-- including Popular Communications, Monitoring Times, club newsletters, and in this book. Manufacturers get wind of what we are doing and conjure up new models that incorporate all of these neat ideas. By then, you have more features you want, and the cycle repeats itself.

Have an idea for some feature you'd like to see in a scanner? Call it to somebody's attention! Don't keep it all to yourself, or else it will never happen-- or else someone else will think of it and become famous while you walk around muttering how you thought of that same thing six months earlier!

I'd like to think that this book will be only the first of many more. Whether or not this is a Series of 1, or 2, or 5, or more depends solely on you, and in two ways. First, you have to buy the book-- and you've already done that. Second, you have to contribute ideas, wish lists, opinions, or the results of your own experimentation. You are cordially invited to send details of your scanner modifications to me and when we get enough of a response, I'll assemble what we have into Volume II. Of course, if there is a tiny response, then it might not happen unless maybe I have a flash of insight and inspiration from UFO's or some Ancient Masters of Wisdom floating around out there on the Cosmic planes. Not that I'm knocking insight-- there was a time (not all that long ago) when everybody was wondering if there were any modifications that could be done to a Realistic PRO-2004 beyond restoring the cellular frequencies. Then it was discovered that the memory could easily be increased from 300 to 400 channels. At that point, the realization came that it was only the beginning! A number of clever hackers were inspired to come up with many more delights.

Now here is my own personal "wish list" for the future. Let's not call it a prediction of things to come since I might be too far out in left field for these to be practical. So, they are my "dream" modifications specifically for the PRO-2004/2005, but the list really applies to any modern, state-of-the-art scanner. I'm not hung up on brand names, just on what the scanner can and cannot do. Accordingly, the PRO-2004/2005 here are representative of a much larger concept where there aren't any brand names (yet).

Wish List for Future PRO-2004/2005 Modifications:

1. Have you ever programmed several hundred channels at a whack. Time consuming and tedious, isn't it? In the era of the computer, where ham radios come with computer interfaces, why not an RS-232 interface for the PRO-2004/2005?

Particularly needed is a simple interface and software so that you can type into a computer a series of channel numbers and frequencies, along with customized settings for mode and delay, then hit a button and have all of the data automatically transferred into the scanner's memory.

2. What about a programmable calculator similar to a Hewlett Packard HP-41CV built in or interfaced to the PRO-2004/2005 for computational capability as well as quick programming and changes in programming?

I know an engineer who built an interface for an HP-41CV to control his ICOM R-71 and R-7000 receivers. Could it be done with the PRO-2004/2005?

3. How about a wideband, low noise preamplifier for the PRO-2004/2005? Hey, don't jump down my throat! No such animal exists, if we're talking about a real "low noise, wideband" preamp! And, I'm talking about a preamp with a bandwidth of 25 to 1300 MHz with a Noise Figure less than 1.0 dB and a gain greater than 10 dB, and one that doesn't go freaky over strong signals.

There are several wideband scanner preamps on the market, but they are using bipolar transistors or MMIC amplifiers which are inherently noisy and don't offer any significant improvement over the total spectrum available to the PRO-2004/2005.

What's called for here is a wideband preamp built around a gallium arsenide field effect transistor (GaAsFET), since it is inherently a low noise device. GaAsFET's have been used for more than a decade in narrow-band, low-noise preamps, but rarely for wideband preamps. You will recall from elsewhere in this book that Gain increases and Noise decreases when bandwidth is narrowed. Conversely, as bandwidth increases, gain drops and noise increases.

Just remember this before you tout what you think is a true low noise, wideband preamplifier, a real one will actually increase the signal-to-noise ratio across the entire band available to the receiver. This is to say that when you hear a weak, noisy signal at 50 MHz and then turn on the preamp, the noise should decrease while the signal volume and clarity should increase. That very same preamp should accomplish the same thing equally at 100, 500, 1000 and 1300 MHz without any retuning or manual control. As I said, it probably doesn't yet exist, but I'd like to hear from the person who claims to have one! If it's for real, I'll try to help him to fame and fortune-- if he or she isn't already wealthy! A "dream" preamp for scannists is outlined in Table 4-24-1.

4. I'm afraid that this one is going to need an external converter, but what about the reception gap from 520 to 760 MHz? Sure, these are UHF TV channels, but recall that only a few years ago the FCC lopped off UHF TV Channels 70 to 83 (806 to 890 MHz) and turned it over to cellular and land mobile communications. Also several large metro areas were told that they could use UHF-TV Channels 14 to 21 (470 to 518 MHz) for two-way radio services (UHF-T band). So who's to say that someday the 520 to 760 MHz spectrum won't have chunks given out to various two-way services? Let's face it, the UHF TV channels are underutilized and each one of those juicy (6 MHz-wide) TV channels can accomodate a large number of 12.5 kHz-spaced NFM two-way radio channels. Many two-way users are screaming for more spectrum-- these frequencies are ripe for plucking-- but scanners don't receive them (yet).

Table 4-24-1

SPECIFICATIONS FOR AN IDEAL SCANNER PREAMP

Noise Figure:	Less than 1 dB over the bandwidth
Gain:	10 dB or greater over the bandwidth
Bandwidth:	25 MHz - 1300 MHz
1 dB Comression:	20 dBm or greater over the bandwidth
2nd Order Intercept:	50 dBm over the bandwidth
3rd Order Intercept:	30 dBm over the bandwidth
Power Requirements:	8 - 16 VDC at less than 50 ma, coaxial fed with DC blocking
Mechanical:	Weatherproof, rugged, mountable at antenna
Other:	Selectable bypass (On/Off or In/Out)
Price:	Less than $150.00

5. Here's yet another tough (but not impossible) one-- how about the built-in capability to receive single sideband (SSB) signals in the PRO-2004/2005? About all it would require is a stable Beat Frequency Oscillator (BFO) externally tunable slightly above and below an IF frequency. A product detector or balanced demodulator would be a better approach, but even a BFO is better than nothing.

6. How about an electronic RF gain control to eliminate that unaccessable "ATT" switch on the rear panel of the PRO-2004/2005? Scanners should have the easy ability to knock down the strength of strong incoming signals. Why Radio Shack decided to stick the "ATT" switch where it's virtually useless, I'll never know. It needs to be on the front panel if it is to be of any value at all. And it would be nice to have an analog control with a range of 20 dB (or more). Better still, would be an attenuation range of 0 to -42 dB that is programmable in 3 dB steps on a per-channel basis like the mode and delay features on the PRO-2004/2005.

7. Let's push for faster scan and search speeds. While 25 to 30 channels or steps per second isn't bad, let's keep in mind that the Uniden/Regency Information Radio scanners have a feature called "Super Turbo Scan II" that races through the channels at 60 to 100 per second! What limits the PRO-2004/2005 to 30 per second? It's apparent that the technology has been developed to double or triple this speed. What can be done to move the PRO-2004/2005 out of that limitation?

8. How about a "Center Tuning Meter" that indicates the precise center frequency of a received signal? Or, at least which direction (high or low) the received frequency actually is? Even a horizontal 5-segment LED would be useful with the center segment to indicate center frequency and the two on either side to perhaps indicate ± 2.5 kHz and ± 5 kHz, respectively.

9. New and improved "S" meters are always something we'd like to see. Even an LED bargraph "S" meter would be better than nothing. I have heard about some LED "s" meters out and around for the PRO-2004/2005, so maybe someone who has a circuit for one will send it along.

10. Here's an idea I have been working on with no success-- an indirect lighting scheme for the keypad of the PRO-2004. It's got to look good as well as being functional. Sure, you can always use a flashlight, or brighter room lighting-- if <u>light</u> is all that is wanted. But it should be more, like maybe a couple of mini-lamps wired inside the front panel with slots cut in the recessed sides of the keyboard rectangle. The critical thing is that direct light from the bulbs should not be visible to the operator. That would run the operator's night vision and also the aesthetic appeal of the radio. Built-in, indirect, low-level illumunation of the keyboard is the objective.

11. What about a spectrum scope on a CRT, like in the new ICOM IC-R9000 shortwave communications receiver? It would indicate all signal activities within a switchable range of frequencies above/below the channel you're monitoring. With one on your scanner, you could monitor one frequency and then visually watch all nearby frequencies for associated activity that might have snuck by you unnoticed.

Write to me. OK, so there are some of the wishes I have, and now I invite you to add to that list. Or, maybe you've heard about a solution to something on this list, or something else new and wonderful for the Realistic PRO-2004/2005 or other modern scanners. If so, please share the information with me. I welcome your correspondence-- and if you include a self-addressed, stamped envelope and one loose extra stamp, your letter will get put near the top of the stack of those to receive a timely reply. Please direct all correspondence to me as follows:

"Doctor Rigormortis," Commtronics Engineering, Communications Department, P.O. Box 262478, San Diego, CA 92126-0969.

MOD-25   If You Can't Do The Work Yourself, or Can't Find The Parts & Materials

This book is designed to make it as easy and inviting as possible for the average user to do the modifications himself or herself. Still, there might be a number of reasons a person can't or doesn't wish to do the modifications in that manner. Or, perhaps a person would very much like to do the modifications, but there is difficulty in rounding up all of the parts, tools, and other materials. Don't fret, there are several options to pursue.

Can't do it yourself? My first recommendation is that you take this book to a competent friend or local communications shop and ask them to do the work for you. There aren't too many professional two-way communications service shops that can do this kind of work profitably, so you might start off by asking around at shops that sell and repair CB sets and scanners. A professional two-way communications shop may produce top quality work, but they'll probably take longer and charge more for the job. But, perhaps you have a friend who can perform the work in a professional manner for a lot less, or for nothing more than the cost of the parts.

If you strike out with these approaches after several tries, my own service shop is available to work with you on a mail-order basis. We have done this type of mail-order technical service since 1975, offering reliable work with a short turnaround time, and at a reasonable cost.

While you can ship your equipment to us at any time, it is recommended that before you ship any equipment, you first write and explain what equipment you have and what work you want done. Give us as much information as you can. We will provide a timely written quotation or estimate along with the latest of any available information relative to your needs. If and when equipment is eventually shipped to us, it is best to use United Parcel Service ground or 2nd day air (depending on urgency), and be sure to insure your shipment at par value.

Commtronics Engineering offers full repair, custom modification, design and engineering services for industry, and for hobbyists in all areas of electronics and communications. We draw a line only in that our work will not turn equipment into anything illegal, nor will we violate any laws or regulations-- please do not ask us to.

Address all inquiries to only the following address: Commtronics Engineering, P.O. Box 262478, San Diego, CA 92126-0969.

If You Can Do The Work But Need Parts, Materials or Kits. By and large, we can't obtain parts and materials any cheaper than you can if you shop around. We do incur considerable extra expense in the purchase, testing, stocking, handling, packaging, and reshipping-- and that must be passed along when people purchase the parts from us.

We aren't in the "parts" business anyway, so it's highly suggested that you attempt to procure the necessities on your own (and we have listed Radio Shack catalog numbers wherever possible to make this convenient for you). But, if you just can't locate needed components, Commtronics Engineering can supply them at cost plus a small margin for expenses and shipping. For MOD-6, MOD-16, and MOD-19, we have preassembled and tested circuit boards available if you don't wish to construct them yourself. Various components are available for most of the modifications shown in this book. Send a list of what you need, and a stamped, self-addressed return envelope with an extra loose stamp and we'll provide you with current availabilities and prices.

Address all inquiries to only the following address: Commtronics Engineering, P.O. Box 262478, San Diego, CA 92126-0969.

No text, diagram, photo, modification, technique, method, procedure, or product in this book should be construed to imply any endorsement whatsoever of this book by any manufacturer of scanning receivers or related products. Neither does the mention or depiction herein of any commercial product or service indicate or imply any connection with, recommendation, or endorsement of those products and/or services or their manufacturing, importing, service, or sales/distribution companies or agents by the author and/or publisher of this book.

Quick-Find Reference Guide

Periodicals

Popular Communications, 76 North Broadway, Hicksville, NY 11801
Monitoring Times, P.O. Box 98, Brasstown, NC 28902
Popular Electronics, 500-B Bi-County Boulevard, Farmingdale, NY 11735
CQ Magazine, 76 North Broadway, Hicksville, NY 11801
73 Magazine, WGE Center, Peterborough, NH 03458

Scanner User Groups

Northeast Scanning News, 212 West Broad Street, Paulsboro, NJ 08066
Radio Monitors Newsletter of Maryland, P.O. Box 394, Hampstead, MD 21074

Publishers of Frequency Directories

CRB Research Books, Inc., P.O. Box 56, Commack, NY 11725
Universal Amateur Radio, 1280 Aida Dr., Reynoldsburg, OH 43068
Gilfer Shortwave, 52 Park Ave., Park Ridge, NJ 07656
Grove Enterprises, P.O. Bpx 98, Brasstown, NC 28902

The Author of This Book

Bill Cheek, Commtronics Engineering, P.O. Box 262478, San Diego, CA 92126

When writing to any of the above, please mention that you got their name from the Scanner Modification Handbook. They'll appreciate it, and so will we! Thanks!

Now There's A Big 220 Page

Scanner Modification Handbook, Volume 2

With 18 More Great Mods-- Complete With Photos

& Easy Step-by-Step Instructions! Volume 2 Also

Contains PRO-2006 Mod Info For All PRO-2004/5

Mods Given in Volume 1!

Ask For It Where You Obtained This Volume--

Or Contact CRB Research If Unavailable There!